Diagnostic Methods in Clinical Thyroidology

J.I. Hamburger
Editor

Diagnostic Methods in Clinical Thyroidology

With 24 Figures

Springer-Verlag
New York Berlin Heidelberg
London Paris Tokyo

Joel I. Hamburger, M.D.
29877 Telegraph Road
Suite 303
Farmbrook Medical Two
Southfield, MI 48034, USA

Library of Congress Cataloging-in-Publication Data
Diagnostic methods in clinical thyroidology / Joel I. Hamburger, editor.
p. cm.
Includes index.
ISBN-13:978-1-4612-8177-1 (alk. paper)
1. Thyroid gland—Diseases—Diagnosis. I. Hamburger, Joel I., 1929–
[DNLM: 1. Biopsy, needle—methods. 2. Thyroid function tests. 3. Thyroid Gland—abnormalities. 4. Thyroid neoplasms—diagnosis. WK 202 D536 1988]
RC655.5.D53 1989
616.4'4075—dc19
DNLM/DLC
88-3864

Printed on acid-free paper

Typeset by Impressions, Inc., Madison, Wisconsin.

9 8 7 6 5 4 3 2 1

ISBN-13:978-1-4612-8177-1 e-ISBN-13:978-1-4612-3612-2
DOI:10.1007/978-1-4612-3612-2

Preface

The evolution of diagnostic methods in clinical thyroidology over the past few decades has been truly remarkable. In my fellowship days the PBI was considered an important advance, rendering the Basal Metabolic Rate (BMR) obsolete. Since then, generation after generation of in vitro tests of thyroid function have come and gone, and we still see exciting new technological advances. An example is immunochemiluminometric assay, a method by which it seems possible to measure serum concentrations of thyroid-stimulating hormone (TSH) and the thyroid hormones without the use of radioactive materials. It is too soon to define the role of this new methodology, but preliminary experiences are included in this volume.

Until recently, who would have dreamed that the first-line in vitro screening test for thyroid function would no longer be a test measuring serum thyroid hormone concentration, but rather one that measures the pituitary hormone TSH? The role in clinical practice of the new supersensitive TSH assays receives considerable attention in this volume. Surprisingly controversial was whether the supersensitive TSH assay was the best method of monitoring thyroxine treatment.

The increasing precision of modern in vitro thyroid function tests has been accompanied by the recognition of a growing array of situations in which euthyroid, or seemingly euthyroid, patients have abnormal thyroid function test values. These include a variety of nonthyroidal illnesses, abnormal protein binding, and drug-induced alterations. Methods to avoid diagnostic pitfalls in these settings are subjects of considerable discussion.

Noninvasive evaluation of structural abnormalities of the thyroid is no longer confined to various imaging techniques. The serum thyroglobulin determination seems destined to play a major role in the follow-up of thyroid cancer patients.

Refinement in thyroid imaging has led to the high-resolution pictures to which we are now accustomed. However, the parallel increase in quality of in vitro studies, and the advent of needle biposy, have resulted in a lessening of the need for thyroid imaging. The continuing indications for imaging are subject to differences of opinion that are explored in this volume.

Needle biopsy has clearly revolutionized the diagnostic evaluation of the thyroid nodule. The issue is no longer whether to use needle biopsy, but rather how to make the most of it. Different attitudes on techniques for fine-needle biopsy and the relative merits of fine- versus coarse-needle biopsy are discussed at some length. The relative uselessness of routine diagnostic ultrasound receives appropriate (in this author's opinion) emphasis.

This book is an outgrowth of a clinical conference held at Sinai Hospital of Detroit, April 20, 1988. Each participant prepared a paper on his assigned subject in advance for review by all of the other participants, who offered questions, comments, and criticisms to which the original authors responded. A principal objective of the seminar was to encourage a free and full interchange of ideas on diagnostic methods. In some cases additional comments were sought from other eminent authorities. It is my belief that the heart of this volume is the exchange of opinions, sometimes rather pointed, always searching and based on extensive practical experience. All controversies were not resolved, but that is the nature of clinical practice. Indeed, the challenge and the frustration of authoring and editing a medical book of this type is knowing when to say enough, even though a definitive resolution of critical issues has not been achieved. In this sense, my book will never be finished.

The topics in this book fall into three general categories: diagnosis of thyroid function, noninvasive evaluation of structural abnormalities, and needle biopsy diagnoses of thyroid nodules. For each category the issues are discussed in depth, bringing into focus areas of controversy. It was my initial idea to leave these controversies for the reader to deal with as best suited his or her style of practice. However, as the organizer of this project, for better or worse, it was my responsibility to resolve clinically relevant controversies within the framework of sound clinical practice. With this objective in mind, and secure in the certain knowledge that I would not fully satisfy any of the participants, I take full responsibility for the summaries that follow each category of topics.

JOEL I. HAMBURGER

Contents

Contributors

MILA BEDNARZ, B.S., Manager/Supervisor, Ligand Laboratory, Office of Dr. Joel Hamburger, Southfield, Michigan 48034, USA

LESLIE J. DEGROOT, M.D., Professor of Medicine, Thyroid Study Unit, Department of Medicine, The University of Chicago, Chicago, Illinois 60637, USA

JOHN T. DUNN, M.D., Professor, Division of Endocrinology, Department of Internal Medicine, University of Virginia School of Medicine, Charlottesville, Virginia 22908, USA

BARRY G. ENGLAND, M.D., Department of Pathology, Ligand Laboratory, University of Michigan, Ann Arbor, Michigan 48109, USA

JOHN E. FREITAS, M.D., Staff Physician, Nuclear Medicine Department, William Beaumont Hospital, Royal Oak, Michigan, 48072, USA; Clinical Associate Professor of Internal Medicine, University of Michigan Medical School, Ann Arbor, Michigan 48109, USA

JOEL I. HAMBURGER, M.D., Southfield, Michigan 48034, USA

IAN D. HAY, M.B., PH.D., F.A.C.P., F.R.C.P., Associate Professor of Medicine, Mayo Medical School; Consultant in Endocrinology and Internal Medicine, Mayo Clinic, Rochester, Minnesota 55905, USA

MUJTABA HUSAIN, M.D., Department of Laboratory Medicine, Sinai Hospital of Detroit, Detroit, Michigan 48235, USA

BUYUNI L. JAHAZI, M.SC.(HONS), M.D., PH.D., Senior Lecturer, Department of Medicine and Department of Biochemistry, The University of Dares Salaam, Dares Salaam, Tanzania; Former I.A.E.A. Fellow, Thyroid Study Unit, Department of Medicine, The University of Chicago, Chicago, Illinois 60637, USA

P. REED LARSEN, M.D., Professor of Medicine, Harvard Medical School, Boston, Massachusetts, 02115, USA

PAUL LO GERFO, M.D., Professor of Surgery, Chief, Division of General Surgery, College of Physicians and Surgeons of Columbia University, New York, New York 10032, USA

AMPICA MANGKLABRUKS, M.D., Clinical and Research Fellow, Thyroid Study Unit, Department of Medicine, The University of Chicago, Chicago, Illinois 60637, USA

MARTIN J. SCHLUMBERGER, M.D., Department of Nuclear Medicine, Gustave Roussy Institute, Villejuif, France

JAMES C. SISSON, M.D., Professor, Department of Internal Medicine, University of Michigan Medical School, Ann Arbor, Michigan 48109, USA

JAMES B. SMART, M.S., PH.D., Senior Research Associate, Department of Pathology, University Hospital, University of Michigan, Ann Arbor, Michigan 48109, USA

ANTHONY D. TOFT, M.D., F.R.C.P.(EDIN), Consultant Physician, Royal Infirmary, Edinburgh; Honorary Senior Lecturer, Department of Medicine, University of Edinburgh, Edinburgh, Scotland

MAURICE R. TUBIANA, M.D., Professor of Experimental and Clinical Radiotherapy, Paris-Sud University Medical School; Director, Gustave Roussy Institute, Villejuif, France

ROBERT VOLPÉ, M.D., F.R.C.P.(C), F.A.C.P., Professor, Department of Medicine, University of Toronto; Director, Division of Endocrinology and Metabolism, University of Toronto; Senior Physician, Department of Medicine, Wellesley Hospital, Toronto, Canada

LEONARD WARTOFSKY, M.D., Professor of Medicine and Physiology, Uniformed Services University of the Health Sciences, Bethesda, Maryland 20814; Chief, Endocrine-Metabolic Service, Walter Reed Army Medical Center, Washington, District of Columbia 20307, USA

Section I
Diagnosis of Thyroid Function

1

Abnormal Thyroid Function Tests in Euthyroid Patients

Leslie J. DeGroot

It may be well to begin this discussion by noting that only one element in the title is unambiguous, and that is "patient." Probably there would be little argument over who is or is not a patient. However, the universe of thyroid function tests is certainly too great to be encompassed in this presentation, and therefore tests of thyroid hormone concentration in blood are the main focus of attention. More importantly, we must recognize that the word "abnormal" could have a variety of meanings. As commonly used it refers to a value outside the range of three standard deviations around the mean of the normal population for that particular test. But clearly, what is an abnormal thyroxine value in a pregnant woman, for example, is quite different from what constitutes an abnormal thyroxine value for a normal adult male. Thus abnormal has to be defined for the particular circumstances. Likewise, euthyroid implies a previously established definition or some independent means of assessing a completely normally functioning thyroid hormone response system, and it will become evident that, in many circumstances, there is no clear way of determining at this time whether "euthyroidism" is present in a given patient or not. With these disclaimers settled, an effort will be made to examine causes of hyper- and hypothyroxinemia and triiodothyroninemia.

Only a few of the many available function tests are considered. Serum thyroxine is usually measured by radioimmunoassay (RIA). The T_3 resin uptake test is used to estimate the free fraction of levothyroxine and is then used to calculate the free thyroxine index (FTI). This widely used test remains satisfactory for most purposes. Newer tests purporting to measure "free T_4" are in general as valuable as the FTI but do not uniformly exceed it in discriminatory ability. Serum T_3 is measured by RIA and is not usually corrected for binding abnormalities, although a free T_3 index can be calculated. Serum reverse T_3 is measured by RIA. Serum thyroid-stimulating hormone (TSH) is best measured by one of the new "supersensitive" assays providing data on elevated, normal, and suppressed values. The gold standard for free T_4 remains the product of total serum T_4 RIA multiplied by the free fraction, measured by dialysis. This is generally a research procedure, and even it may produce values of uncertain meaning in some clinical settings.

The proteins that can bind—or carry—thyroid hormone and triiodothyronine in blood are thyroxine-binding globulin (TBG), thyroxine-binding prealbumin, albumin, and in abnormal circumstances immunoglobulins. Congenital (inherited) increase in TBG occurs in about 1 per 20,000 individuals as an X-linked trait, causing an increased quantity of normal TBG.[1] This relatively common abnormality, fully expressed in males and less fully in females, causes up to a threefold elevation of T_4 and to a lesser extent elevated T_3 levels, but does not affect supply or utilization of thyroid hormone in any way and is fundamentally only a cause of confusion in diagnosis. Almost all current methods of estimating free thyroid hormone levels, including calculation of a free thyroxine index from total thyroxine and resin uptake values, or newer estimates of "free thyroxine," adequately correct for this variation and indicate in these patients a normal supply of free hormone. Table 1.1 lists causes of hyperthyroxinemia due to serum protein binding abnormalities. As indicated in Table 1.2, a direct immunoassay of TBG proves the diagnosis, but it is usually possible to assume the diagnosis on the basis of elevated T_4, mildly elevated T_3, reduced resin uptake test, a normal metabolic status, normal TSH, and a similar pattern in other family members.

Thyroxine-binding prealbumin abnormalities as a cause of elevated T_4 have been reported rarely,[2] and it is not possible to predict with certainty

TABLE 1.1. Hyperthyroxinemia due to serum protein binding abnormalities

Congenital
↑ TBG
↑ TBPA
FDH
Acquired
Newborn
Pregnancy
Nonthyroidal disease
Acute illness
Hepatitis and hepatoma
Estrogen-producing tumor
Hydatidiform mole
Porphyria
Drugs
Estrogen
Heroin
Methadone
5-Fluorouracil
Clofibrate
Perphenazine
Antibodies to T_4

TABLE 1.2. Differential Diagnosis of binding abnormalities in euthyroid patients.

	TBG excess	Familial dysalbuminemic hyperthyroxinemia	Prealbumin associated hyperthyroxinemia	T_4 binding antibodies
	X-linked	Autosomal dominant	?	Usually sporadic (associated with thyroiditis)
T_4	↑	↑	↑	↑ Normal or ↓ (method dependent)
RT_3 Uptake	↓	Normal	Normal	Variable
RT_4 Uptake	↓	↓	?	Variable
FTI	Normal usually	↑	↑	Variable
T_3	↑	Normal or slightly ↑	Normal	Variable
Binding characteristic	$T_4 + T_3$	T_4, not T_3	T_4, not T_3	May bind T_4 or T_3 or both
Dialyzable free T_4	Normal	Normal in absence of Cl− ion	Normal	Normal
Differential test	T_3RU or TBG measurement	Low charcoal uptake or resin uptake with T_4 load; Immunoprecipitation of T_4 by anti-albumin	Immunoprecipitation of T_4 by anti-TBPA; Electrophoresis	Excess T_4 or T_3 precipitated by PEG

how they would affect assessment of free thyroid hormone by various methods. The most common presentation would be with elevated T_4 and normal resin uptake, giving an elevated free thyroxine index in a patient without thyroid disease and with a normal TSH and serum T_3. Free T_4 by dialysis should be normal.

One of the more interesting recently detected causes of elevated T_4 levels is the syndrome of "familial dysalbuminemic hyperthyroxinemia"(FDH),[3,4] in which patients inherit, in an autosomal dominant manner, an albumin that binds thyroid hormone more tightly than normal and thus causes an elevation of total thyroxine in blood. The affinity of this albumin, or group of albumins, is much less for triiodothyronine, and therefore, serum T_3 values are usually normal or only minimally elevated. This is often a tip-off to the diagnosis. Because this protein does not bind triiodothyronine actively, T_3 resin uptake tests do not provide a correction for the abnormal binding, and in such patients, elevated FTI values suggest hyperthyroidism. Resin uptake tests done using labeled thyroxine automatically correct for the defect and give a normal interpretation of the metabolic status, as do assays of free T_4. Again a tip-off to this diagnosis is often the occurrence of similar abnormalities in family members.

Acquired hyperthyroxinemia due to protein binding abnormalities are primarily related to estrogen exposure and occur in newborn infants because of maternal estrogen (lasting for 2 to 4 weeks after birth), in women taking oral contraceptive pills, and during pregnancy. TBG elevation also occurs in some patients with hepatitis, in the rare individuals with estrogen-producing tumors, sometimes in association with hydatidiform mole, and in porphyria. Several drugs, including, of course, estrogen, but surprisingly heparin, methadone, 5-fluorouracil, and clofibrate, have been reported to increase TBG.[5,6] Usually the discrepancy between the thyroxine elevation and the clinical status, the concomitant elevation of T_4 and T_3, the clinical circumstance, normal TSH and "correction" to a normal value with the use of a free thyroxine index, adequately serves to identify patients with acquired elevation of TBG.

Antibodies directed against thyroxine and/or triiodothyronine occur occasionally in individuals with autoimmune thyroid disease. These antibodies may bind large quantities of thyroid hormone in serum but do not alter the metabolic status.[7] The effect on thyroid tests can be quite variable, depending on the method used to separate bound from free hormone in the RIA. Levels of serum T_4 or T_3 may be elevated, leading to a suspicion of an unusual metabolic state, since the patients are clinically euthyroid and have normal TSH values.

Differential diagnosis of binding abnormalities is not imperative in, for example, TBG excess, when the FTI or free T_4 assay gives a normal value. However, the situation gets more complicated in patients with FDH, abnormal prealbumin, or thyroxine-binding antibodies. Isoelectric focusing is available in a few laboratories and will demonstrate binding to the specific type of protein and offers a conclusive diagnosis. Often the nature of the unusual binding protein can be determined by immunoprecipitation (Table 1.2). For example, TBG can be measured by immunoassay, the abnormal T_4 binding albumin in FDH can be detected by immunoprecipitation with antialbumin, the abnormal prealbumin can be detected by electrophoresis, and binding of thyroxine to specific immunoglobulins (IgGs) can be detected by precipitation of the IgG using polyethylene glycol or a "second antibody" system.[8]

Causes of euthyroid hyperthyroxinemia without serum protein binding abnormalities are listed in Table 1.3. Elevation of thyroxine, typically with low T_3 and normal TSH, can occur occasionally in seriously ill individuals, although the more common phenomenon is hypothyroxi-

TABLE 1.3. Euthyroid hyperthyroxinemia without serum protein

Binding Abnormalities
Illness associated
Acute illness
Psychiatric disorders
Hyperemesis gravidarum
Drug related
Propranolol
Ipodate
Amiodarone
Thyroxine related
During T_4 therapy
At onset of therapy
2–4 hours after T_4 ingestion
High altitude
Thyroid hormone resistance
Abnormal $T_4 \longrightarrow T_3$ conversion

nemia. Possibly this is a transient effect of diminished 5′-T_4 deiodinase activity during acute illness.[9] Elevated T_4 values, FTI values, and usually normal T_3 values have been reported with acute psychiatric illness, with a tendency for these abnormalities to self-correct over a period of 4 to 8 weeks.[10] TSH may be normal, but thyrotropin-releasing hormone (TRH) response is often blunted. The cause of these phenomena is unknown but may be related to prior use of sympathomimetic drugs that could affect TSH or thyroid gland stimulation directly, or represent some kind of "activation" of the hypothalamic-pituitary-thyroid axis. The matter is unresolved. At a practical level, it is inappropriate to diagnose hyperthyroidism in recently admitted patients with serious psychiatric illness, especially in the absence of thyroid enlargement, suppressed TSH, or autoantibodies. However, in truth, whether this represents a state of hyperthyroxinemia with euthyroidism, or hyperthyroidism, is uncertain.

Hyperemesis gravidarum is also often associated with hyperthyroxinemia, mildly elevated T_3 values, possibly normal TSH values, and flat TRH responses.[11] Again, the mechanism is unknown. Individuals who are suddenly exposed to high altitude may develop elevated T_4, elevated FTI, elevated T_3, and elevated TSH.[12] It has been speculated in these circumstances that some activation of the hypothalamic-pituitary-thyroid (HPT) axis has occurred, and it is possible that this actually represents a transient, mild state of hyperthyroidism without obvious clinical signs.

Commonly used drugs, including propranolol, ipodate, and amiodarone, may cause thyroxine elevation, typically in association with depressed serum T_3 levels. These drugs inhibit 5′-T_4 monodeiodinase, thus seeming to "dam up" the metabolism of thyroxine, causing its elevation.[13] In some instances, elevated T_4 may actually be due to an augmented output of thyroxine because the pituitary is not receiving a sufficiently active suppressive signal from intrapituitary-generated triiodothyronine, due to the inhibition of 5′-deiodinase. The individuals appear to be euthyroid and have normal TSH levels.

Amiodarone (Table 1.4) is a drug with a variety of effects on the HPT axis and can cause many problems in diagnosis.[14,15] By inhibiting T_4 to T_3 conversion, it may increase T_4 and FTI and diminish T_3, and by diminishing clearance of reverse T_3, it causes elevation of this compound in blood. It may induce hypothyroidism, probably by the release of iodide from its molecule, producing an inhibitory effect on thyroid hormone synthesis and release. This occurs in patients with and without Hashimoto's thyroiditis. The drug is also associated with thyrotoxicosis, frequently occurring after amiodarone has been discontinued, probably in some way due to the huge body-burden of iodine, and in this instance is associated with elevated serum T_3 levels and suppressed TSH. This typically occurs in patients with multinodular goiters in areas of relative iodine deficiency.

Hyperthyroxinemia is a very common, if not typical, feature of thyroxine replacement therapy. It occurs in millions of patients who are by other tests deemed euthyroid. Using conventional doses of L-thyroxine such as 0.1 to 0.15 mg/day for replacement or suppressive therapy, patients will be found to have serum T_4 and FTI levels that are on average about 20% to 30% above the normal mean, thus extending well into the low hyperthyroid range. About 30% of these patients have elevated values. Such patients typically will have normal T_3 and TSH levels. Presumably this is partially due to the fact that thyroxine pills contain only T_4, whereas the normal thyroid secretes T_4 and T_3. Thus a somewhat higher than normal level of T_4 is required to generate the appropriate levels of T_3 to maintain euthyroidism. While it is easy in practice to automatically compensate for this change in the normal range of FTI values for patients on replacement therapy, it is probably more practical nowadays to rely on the "supersensitive" TSH assays to determine whether patients are euthyroid or hyperthyroid. Current evidence supports the idea that maintenance doses of thyroxine that keep the patient's TSH in the normal

TABLE 1.4. Effects of amiodarone.

1. Decreased T_4 to T_3 conversion
   Thus normal or ↑ T_4 and FTI, ↓ T_3
   Decreased rT_3 clearance, ↑ rT_3*
2. Hypothyroidism (due to excess I_2) and ↑ TSH in some patients
   Not specifically associated with Hashimoto's thyroiditis
3. Thyrotoxicosis, especially after drug withdrawal
   T_3 ↑
   Probably associated with multinodular goiter

*Reverse T_3.

range are probably providing a normal metabolic status for the patient's whole body.[16] Obviously one is in fact measuring the apparent metabolic status of the pituitary alone, but it is also probable that the metabolic status of other organs more or less falls in line with that of the pituitary. T_4 tends to be especially elevated in patients during the institution of thyroid hormone therapy on any given dosage, probably because it takes some time for the thyroxine-metabolizing enzymes to become normal after being suppressed during hypothyroidism.[17] Thus, at the institution of a given level of T_4 replacement, thyroxine will be a bit elevated, but after 4 to 6 seeks, will achieve an appropriate lower steady state level. Another anomaly can occur each day after the daily dosage of thyroxine is given, which causes an elevation in serum T_4 and FTI. Individuals receiving more than 100 μg/day will have up to 1 to 2 μg% elevation in T_4 occurring 3 to 6 hours after ingestion of the dose.[18]

Elevation of thyroid hormone levels due to resistance to the metabolic action of thyroid hormone is a topic of considerable current interest. Three syndromes have been described (Table 1.5). In patients with generalized resistance to thyroid hormone (Table 1.6), the individuals may appear clinically euthyroid or hypothyroid; typically have a goiter; infrequently have retarded bone development or hearing defect; have normal serum markers of thyroid hormone activity such as cholesterol, carotene,

TABLE 1.5. Resistance to thyroid hormone.

Generalized resistance
Pituitary resistance
Elevated T_4 and T_3 and normal TSH
Thyrotoxicosis
No pituitary tumor
Peripheral resistance to thyroid hormone

TABLE 1.6. Generalized resistance to thyroid hormone.

Clinically euthyroid
Goiter
Variable normal or retarded bone age
Infrequent hearing defects
Tachycardia
Elevated free T_4, T_3, rT_3,* thyroglobulin
Normal or increased basal TSH and TRH response
Absent thyroglobulin and microsomal antibodies
T_4 and T_3 turnover augmented
BMR, cholesterol, carotene, sex-hormone-binding globulin normal
Resistance of TSH to T_4 and T_3 suppression
Reduced tissue chemical responses to T_4 and T_3
Normal TSH suppression by steroids

*Reverse T_3

hormone binding globulin levels; have normal or elevated TSH; respond to TRH despite elevated total and free T_4 and T_3 levels; and have, in the steady state and when untreated, augmented T_4 and T_3 turnover rates.[19] Their pituitary is resistant to T_3 suppression. This can be demonstrated by serial TRH studies during progressive increments of administered T_3. Normal individuals usually have a suppressed TRH response with 50 to 100 μg of triiodothyronine qd, whereas in resistant individuals, it may take up to 300 μg/day to completely suppress TRH responsiveness. In contrast to resistance to thyroid hormone, steroid administration prevents TRH stimulation of TSH secretion in the usual manner.[20]

In the syndrome of pituitary resistance to thyroid hormone, because hormone cannot normally suppress TSH production, the pituitary continues to secrete TSH, driving up serum T_4 and T_3, and causing thyrotoxicosis.[21] This condition must be differentiated from the equally rare occurrence of pituitary tumors secreting TSH and causing hyperthyroidism, which is associated with increased production of free TSH alpha subunits[22] and usually is unresponsive to stimulation with TRH or suppression by supraphysiologic doses of thyroid hormone.

Isolated peripheral resistance to thyroid hormone has also been reported, but the definition of this syndrome remains unsatisfactory.

Decreased thyroid hormone in serum, in the presence of an apparent normal metabolic state, occurs commonly due to inherited defects in binding serum proteins, or because of drug effects (Table 1.7). Serious illness is another major cause of this condition and is discussed subsequently.[7]

Inherited TBG deficiency is detected in about 1:5,000 neonates. Because the condition is X-chromosome linked, detected newborns are almost exclusively males. The overall incidence including heterozygous females is about 1:2,500. Some individuals, for example, Australian aborgines, inherit a defective TBG that also clinically presents with low hormone levels.[22] Abnormalities in TBG are usually easily understood because of the "correction" using the Free Thyroxine Index, or more specifically

TABLE 1.7. Euthyroid hypothyroxinemia.

Decreased thyroid hormone binding
Inherited
Decreased thyroxine-binding globulin
Abnormal thyroxine-binding globulin
Acquired
Decreased thyroxine-binding globulin
Decreased prealbumin
Drug related (Dilantin, ASA, phenylbutazone, testosterone, fenclofenec, mitotane, L-asparaginase)
Nonthyroidal illness
Triiodothyronine (T_3) therapy

diagnosed by direct measurement of TBG levels using immunoassay techniques, the normal metabolic status, lack of goiter, normal TSH levels, and familial pattern for the condition.

Acquired diminution in TBG occurs in individuals with serious liver disease, nephrotic syndrome, and during administration of testosterone and high doses of glucocorticoids.[23] Decreased prealbumin occurs frequently in illness, but because it contributes little to hormone transport, it has minimal effect on T_4 level.

Several drugs in common use have clinically very important effects on thyroid hormone levels in blood. Of these the most important are Dilantin and aspirin. Aspirin competes with T_4 for binding to TBG causing a low T_4, and in many indirect methods for assaying free hormone, leads to an apparently low value.[24,25] Aspirin has other effects on thyroid hormone economy, but probably the inhibition of binding is the most important. Dilantin also inhibits binding of T_4 to TBG and thus lowers T_4. Again, the commonly used resin uptake methods for measuring free thyroxine indirectly do not "correct" for the effect of Dilantin, but consideration of the whole clinical picture, recognition of ingestion of more than 200mg of Dilantin daily, and lack of goiter and normal TSH usually allows one to make the proper diagnosis. Dilantin has several other effects on thyroid hormone economy, including inhibition of hormone absorption and augmented metabolism of thyroid hormone, which doubtless contribute to the serum abnormalities, although inhibition of binding is probably an important phenomenon. Lastly, administration of triiodothyronine suppresses thyroid gland activity, reduces thyroid hormone secretion, and in proper dosage leaves patients metabolically normal with a normal TSH. Observation of a low T_4 in this setting can lead to the erroneous assumption that the patient is not receiving enough hormone, thus causing the physician to increase the dose of triiodothyronine. It is imperative that physicians understand the difference between the "T_3 test," referring to a resin uptake test, and "T_3" or "serum T_3," or "serum T_3 by RIA," the assay for serum triiodothyronine by radioimmune methods. In the circumstances described with administration of T_3, the T_3 resin uptake test would be typically in the low range, but serum triiodothyronine values are normally very elevated during administration of even physiologically replacement doses of the thyroid hormone. Fortunately, T_3 is used infrequently for therapy of hypothyroidism, since it certainly is prone to cause confusion in diagnosis and treatment.

Lithium chloride should perhaps be added to the list of drugs causing a hypothyroxinemic state, although not euthyroidism.[26] Lithium blocks several actions in the normal pathway of thyroid hormone production and release, probably most importantly the cyclic-AMP-mediated release of thyroid hormone. This is especially evident in individuals who already have an abnormality in thyroid hormone biosynthesis such as in Hashimoto's thyroiditis. Lithium therapy in normal individuals may simply cause a minimal decrease in T_4 with modest elevation of TSH, but in individuals with some susceptibility, such as thyroiditis, it may induce significant hypothyroidism and require replacement therapy. It often causes enlargement of previously present multinodular goiters.

Low Triiodothyronine Syndrome

This syndrome and the companion but more fully developed condition usually described as the "euthyroid sick syndrome" are discussed here briefly and more fully covered elsewhere. The most common cause of hypotriiodothyroninemia, doubtless exceeding the incidence of true hypothyroidism, is the dramatic reduction of triiodothyronine observed in many clinical situations including in the fetus and newborn infant, in elderly individuals, in all starved or sick patients, and in patients taking cortisone, propranolol, opodate, or amiodarone.[27] Thus almost every patient seriously ill in the hospital will have a low serum T_3. These changes are universally due to a reduction in the activity of the 5′-thyroxine monodeiodinase present in virtually all tissues, which provides most of the circulating serum T_3. During starvation this decrease in deiodination is probably due to a reduction in carbohydrate intake, somehow leading to a reduction in generation of reducing equivalents and thus altering the function of the enzyme. Many of the conditions described have in common the feature of starvation. Drugs such as amiodarone, opodate, and propranolol presumably directly inhibit enzyme activity. These patients have a low serum T_3, typically a normal TSH and TRH responsiveness, and classically an elevated reverse T_3, because of inhibition of metabolism of this iodothyronine through the same pathway, and judging from animal studies and minimal human data, would have low tissue triiodothyronine and reduced saturation of thyroid hormone nuclear receptors. Pituitary unresponsiveness to low serum T_3 is probably because the pituitary senses locally generated T_3 formed from T_4 by a second 5′-iodothyronine monodeiodinase characteristically present in pituitary tissue. The metabolic status of the peripheral tissues is not certain. A commonly accepted hypothesis is that this represents a "good" metabolic adaptation designed

TABLE 1.8. Low T_3 syndrome.

Caused by disease, drugs, starvation
↓ T_3, normal TSH, normal TRH response, rT_3* often ↑
Tissue T_3 probably low
Pituitary unresponsiveness unexplained (possibly serum T_4 controls it)
Confusion with hypothyroidism (low rT_3)
Possible cause of normal T_3 in thyrotoxicosis

*Reverse T_3.

to conserve energy stores from destruction by "catabolic" effects of thyroid hormone during periods of metabolic stress. Such a teleologic explanation is absolutely without foundation. Furthermore, animal models of conditions associated with the low T_3 syndrome suggest the presence of hypothyroidism at the level of peripheral tissue. On the other hand, this metabolic abnormality does not appear in itself to be associated with serious adverse consequences in the host, despite what is probably some degree of tissue hypothyroidism. Restoration of nutrient intake, or elimination of stress, illness, or drugs, causes rapid reversion of the serum and presumably tissue T_3 levels to normal.

Patients who develop the "low T_3 syndrome" when previously thyrotoxic can have serum T_3 levels lowered to the normal range during this metabolic state. Again, whether this represents transient metabolic normality or not is uncertain, but the most common answer would probably be that this represents a variety of temporary remission of hyperthyroidism.

Euthyroid Sick Syndrome with Low T_4

The euthyroid sick syndrome with both depressed T_3 and T_4 represents the more severe extension of the "low T_3" syndrome. Patients in the same categories described above with severe illness and starvation, thus representing almost all patients in a medical intensive care unit, have not only a low T_3 but also a low serum T_4. The extent of drop in the T_4 can be severe, and the degree of the drop correlates with the probability that the patient will die of the illness.[28] In such patients, serum free T_4 has been assessed by a variety of methods, and the results are variable, ranging from below normal, to normal, to elevated.[29] Such variations are thought to be due to the presence of substances in the blood of seriously ill patients

TABLE 1.9. Low T_4 syndrome.

Worse than the low T_3 syndrome
T_4 ↓, T_3 ↓, TSH normal, TRH response variable
rT_3* not reliably increased
FTI by conventional methods typically low
TSH and even TRH not totally reliable
In some cases,
Free T_4 by dialysis ↑ or normal
"Free T_4" normal
Possible dialyzable T_4-binding inhibitor
In other cases,
Serum and tissue T_3 and T_4 probably low
Pituitary unresponsiveness
Diagnosis not certain but free T_4, "free T_4," and rT_3 probably most useful
Disappears with recovery

*Reverse T_3.

that inhibit binding of T_4 to TBG and possibly to tissue-binding proteins.[30] These substances could include an inhibitor of phagocytosis,[31] unusual serum proteins, free fatty acids[32] and even therapeutic agents such as Lasix. The low T_3 is also ascribed to the inhibition of 5′-deiodinase activity. Typically patients have a low or normal TSH and respond normally or subnormally to TRH. Because of the low serum T_4 substrate supply, reverse T_3 may not be elevated. Whether the patients are or are not hypothyroid is hotly debated. Surprisingly enough, there is very little data on tissue T_4 and T_3 levels in such patients.[33] Almost certainly, nuclear thyroid hormone receptor levels are reduced.

One school of thought holds that the free T_4 values are actually normal, since the free fraction is elevated despite the data provided by FTI, and that the patient should be left without therapy.[34] A second school of thought concedes that the low serum T_4 and T_3 values probably are associated with low tissue hormone levels, and that the patients are metabolically hypothyroid, but that this is an advantageous metabolic adaptation that should not be treated. A third school of thought suggests that the patients are hypothyroid, as determined from in vitro study and experimental animal models, and that replacement therapy would be appropriate. In this construction, the basic defect is thought to reside either in the hypothalamus, with inadequate generation of TRH leading to inadequate TSH responsiveness to drive the thyroid to increase serum T_4 levels, or the pituitary, which generates relatively larger amounts of the T_3 through increased T_4 deiodination and prevents TSH increase. That this is probably true, in some cases, is indicated by 1) their normal response to TRH, 2) because hypothyroid individuals who develop this syndrome tend to have lower TSH values during the "euthyroid sick" period,[35] and 3) in many individuals with this illness, TSH rises transiently above the normal value during recovery.[36] Another reason to favor the view that the patients are hypothyroid is that, in many patients, in my experience, administration of replacement doses of thyroid hormone restores serum T_4 to euthyroid levels.

Interpretation of the meaning of the serum hormone findings is of course crucially related to therapeutic intervention. The most common approach is not to treat such patients, although this author believes that it is more appropriate to provide replacement doses of T_4 and T_3. What little data are available are inconclusive in establishing that this is beneficial. Some studies have suggested benefit and others none.[37,38] The only really outstanding study conducted in this field so far did not establish the benefit of replacement therapy in a small group of carefully studied patients and controls.[39] Probably if therapy is to be given, it is logical to give a mixture of T_4 and T_3 in a manner that gains direct access to the bloodstream. Also, in the patients who have this illness, it is obvious that a whole variety of organ systems are undergoing degradation and that treatment of thyroid hormone and maintenance of fluid and nutritional support represent only a few of the many ways of supporting individuals with a variety of yet undisclosed or even untreatable systemic abnormalities.

Discussion by Dr. James C. Sisson

You described the syndrome of familial dysalbuminemic hyperthyroxinemia (FDH) as manifesting elevated FTI values since the T_3 resin test does not correct for the binding abnormality. In Table 1.2, the differential tests for this type of hyperthyroxinemia (charcoal uptake, T_4 loaded resin uptake, and immunoprecipitation of T_4) are not readily available to the practitioner. As noted, under special conditions the dialysis method gives a normal free T_4 concentration. What are the results of the commercially available direct assays for free T_4 in patients with the FDH syndrome?

One mechanism by which administered thyroxine induces hyperthyroxinemia is, in the absence of triiodothyronine secretion, more thyroxine must be present to generate triiodothyronine through deiodination. However, the secreted T_3 constitutes only a small fraction of the total T_3 generated, and compensation by added thyroxine should then be commensurately small. Possibly the route of entry of thyroxine into the body plays a role. Since ingested thyroxine enters via the portal vein system, the liver is exposed to unphysiologically high levels of thyroxine. Could the deiodinase enzymes in the liver then be partly inhibited, and more thyroxine then be necessary to generate triiodothyronine?

Response by Dr. Leslie J. DeGroot

I have not systemically reviewed the free T_4 assay results in FDH syndrome. Although the suggestion that thyroxine could inhibit the deiodinase because of high concentrations in liver might make sense during the "first pass" situation, I believe the whole thyroxine-T_3 system equilibrates within minutes to hours, and thus I think this explanation seems highly unlikely to be important in the overall metabolism of T_4 to T_3.

Discussion by Dr. Ian D. Hay

Despite his initial misgivings regarding the definitions of "abnormal" and "euthyroid," Dr. DeGroot has done an excellent job of defining causes of hyper- and hypothyroxinemia and triiodothyroninemia. There are several areas in his survey that, I believe, have in previous reviews received inadequate emphasis. Among these, I would mention the differential tests described in Table 1.2 and the hyperthyroxinemia found at the onset of T_4 therapy and within 2 to 4 hours after ingestion.

I absolutely agree with the use of "supersensitive" TSH assays in the monitoring of T_4 replacement therapy.

If we are ever to overcome the widespread misunderstanding that continues to exist surrounding the "T_3 resin uptake test," I believe that we, as thyroidologists, should start to use for this test the term *thyroid hormone binding ratio* (THBR), which was defined in the recent report of the American Thyroid Association Nomenclature Committee.

Discussion by Dr. John T. Dunn

You have said that a decreased carbohydrate intake decreases the reducing equivalence and enzyme function in starvation. Can you give more details and references for this comment?

Response by Dr. Leslie J. DeGroot

The reference I would give is to TJ Visser, Mechanism of action of iodothyronine 5′-deiodinase, in Biochem Biophys Acta 1979; 569:302–308. Thiols are an obligatory cofactor in the reaction, probably keeping the deiodinase in the reduced state. The natural cofactor is unknown, although reduced glutathione may be important. Metabolism of carbohydrates would serve to keep glutathione in the reduced state.

Discussion by Dr. Joel I. Hamburger

The clinical Assays Gammacoat two-step free T_4 method gives reliable data in the FDH syndrome. We keep a supply of serum from a family with FDH to test the reliability of Free T_4 kits, and so far none of the others we have tested have been reliable.

You said that T_4 values rise 1 to 2 μg% 3 to 6 hours after the ingestion of more than 100 μg of thyroxine. However, Saberi and Utiger[40] showed that there were no such acute changes in either T_4, T_3, or TSH levels. Do you have any further information on that point?

Response by Dr. Leslie J. DeGroot

In addition to the report of Ingbar et al[18] there are two papers that say the same thing, one by Symons and Murphy[41] and a similar study published by Anders Wennlund.[42] I think the study by Utiger just missed the increase because of the timing of samples.

Response by Dr. Joel I. Hamburger

I agree with you, based not only upon your references, but on a study of myself and eight other volunteers from my staff who took Levothroid, 0.15 mg, and had T_4 values obtained 2, 4, and 6 hours after baseline values. Just as you said, the T_4 values increased 1 to 2 μg/dl, at 2 and 4 hours after ingestion of the medication. Perhaps Utiger's data were flawed by the imprecise and now obsolete competitive protein binding assay for T_4. This information suggests that it would be preferable to have patients avoid taking their thyroxine tablets on the day they come for a serum thyroxine determination.

Discussion by Dr. Leonard Wartofsky

Don't you think it might be imprudent to advise that the low T_3 or low T_4 syndromes of critical illness should be treated with combination T_3/T_4 in the absence of any believable data indicating that such therapy may be beneficial and in the presence of data to the contrary (the clinical study you cite by Brent and Hershman as well as other clinical and experimental studies)?

Commentary by Dr. Leslie J. DeGroot and Dr. Ampica Mangklabruks

Our comments will be confined to the clinical problem of patients with serious nonthyroidal illness (NTI) and serum thyroxine values below 5 μg/dl. Patients with illness-associated diminished serum T_3 values and normal T_4 do not appear to be at greatly increased risk, and indeed this finding is extremely common in ill or even briefly starved patients. Patients with illness and serum T_4 values below 5 may be either hypothyroid or not hypothyroid, but their prognosis as a group is bad. In fact, in this setting, there is an inverse correlation between the depression of the serum thyroxine level and possibility of recovery, such that death is the typical outcome of such patients whose thyroxine falls to levels of 1 to 2 μg/dl. Serum TSH levels, in patients with illness-associated depression in T_4 and T_3, are normal, occasionally depressed or elevated, and TRH responsiveness may be reduced. In this setting, an elevation of TSH generally is considered to indicate primary thyroid failure and is a signal for replacement therapy given in a dosage appropriate for the patient's age, evidence of hypothyroidism, and associated complicating illnesses such as cardiovascular disease. We note that elevated TSH is occasionally found in patients whose final diagnosis is NTI, not primary thyroid failure. Controversy does not surround management of patients whose TSH is elevated and who presumably have bona fide hypothyroidism, but rather those with the classic NTI-associated low serum T_4, and in whom the diagnosis of hypothyroidism is not clear by laboratory or clinical means.

The first important question is the true value of the free thyroid hormone in such patients. Unfortunately, as shown in Table 1.10, (27,43–51) different assays provide conflicting results. We have compiled data from the available studies that compared assay methods in groups of patients defined as having nonthyroidal illness and serum thyroxine values below 5. Most patients had FTI values that were low. Free T_4 mea-

TABLE 1.10. Thyroid hormone assays in patients with NTI and $T_4 < 5$.

Reference	Number of cases	FT_4 Dialysis, RIA	FT_4 Index	FT_4-TBG (TBG RIA) Corning	FT_4 "Gammacoat" clinical assays	FT_4 "Immophase" Corning (Method 1)	FT_4 "Liquisol" Damon	FT_4 "Immophase" Corning (Method 2)	FT_4 Amerlex	rT_3*	T_3 Resin uptake ratio (A)	FT_4 Index B Bermudez method	FT_4 Abbott "Tetrazyme"	Ultra-filtration
			(Talc)											
Melmed et al[43]	14	43%↓	71%↓	43%↓	28%	100%↓	29%↓			64%↑				
Chopra et al[44]	11	55%↑	64%↓			18%↑					64%↑	55%↓		
			(Charcoal)											
Kaplan et al[45]	17	18%↓	53%↓											
Wood et al[46]	42		43%↓											
Kaptein et al[47]	6		100%↓							100%↑	100%↑			
Chopra et al[27]	18	44%↑ 17%↓	72%↓							61%↑		56%↓		
Kaptein et al[48]	16	13%↑ 19%↓	100%↓	69%	44%↓ 13%↑	94%↓	87%↓					81%↓	13%↑ 13%↑	
Slag et al[49]	20	45%↓	30%↓	35%	15%↑ 15%↓	70%↓	10%↑ 0%↓					25%↓		
Wang et al[50] (Mean T_4 = 5.42)	20	100%N	Mean ↓36%						35%↓					5%↓
Braverman et al[51]	17	100%N	53%↓	41%		6%↓								

Note. Values approximated from published results.
*Reverse T_3.

sured by the Corning Immunophase method was low in 6% to 100% of patients, and was low in 0% to 87% using the Damon Liquisol method. In contrast, assays measuring free T_4 by dialysis or using the Clinical Assays Gammacoat system have a small proportion of values that fell below normal (0% to 47% and 13% to 28%). But coincidentally, with these assay systems, 13% to 55% of patients using dialysis and 15% to 44% using Gammacoat have elevated free thyroxine values determined by the same methods. It is by no means clear which method provides the "correct" value. TSH values in these patients are usually normal. Reverse T_3 levels may be elevated or normal and are rarely low, but typically are not elevated in patients with clear-cut hypothyroidism associated with elevated TSH levels. TSH values tend to become transiently elevated in NTI patients who recover during the first month after recovery from illness, and this has been taken as a sign of prior hypothyroidism.

Patients with NTI are typically seriously and frequently terminally ill. Often they are on a variety of medications including dopamine, steroids, and heparin. They are invariably starving. Multiple organ systems, including liver, kidney, cardiovascular system, and lungs are failing. Thus it becomes an important therapeutic question as to whether thyroid hormone therapy is 1) of value, 2) not harmful but not helpful, or 3) disadvantageous.

Various interpretations have been offered for the metabolic status of these patients:

1. One hypothesis is that the true free thyroxine is actually normal, and the problem is basically a test abnormality caused by an alteration in serum factors that inhibit binding of thyroxine. According to this assumption, if the proper test was available, the problem would vanish. From the compilation of data provided in Table 1.10, it is not possible to indicate that one test is clearly superior to another, or in fact that any one test gives a true assessment of free hormone values in all patients. While measurement of free thyroxine by dialysis is generally accepted as the benchmark method in well patients, it is by no means obvious that an assay system that indicates that up to 50% of patients may have elevated hormone levels, and is radically contradicted by a variety of other hormone assays, does in fact give a correct value in patients with this illness. It is perhaps most certain that we do not have a proper yardstick against which to compare methods as of this time.
2. A second explanation of the pathophysiology of NTI is that an abnormality affecting binding of thyroxine is present, and active both in serum and tissues. Much evidence for the presence of such a binding inhibitor factor has been provided by Chopra and associates.[30] Various substances including proteins and free fatty acids are considered to be important. However, if one assumes that the important factor inhibits binding of thyroxine, and presumably T_3, to receptors, it must be as-

sumed until proven otherwise that the occupied receptor level is reduced, and the patients could be metabolically hypothyroid.

3. A third explanation for the pathophysiology is that NTI is a form of hypothalamic hypothyroidism due to diminished production of TRH or possibly diminished pituitary responsitivity to normal levels of TRH. If, for example, the pituitary is actually exposed to a normal level of intracellular T_3 because of increased function of 5′-iodothyronine monoiododeiodinase, and augmented conversion of T_4 to T_3 in the pituitary, the pituitary could be "euthyroid," while the rest of the body is hypothyroid. Or, if TRH production is suppressed by the effect of illness, TSH secretion would be diminished in relation to the apparent low T_4 levels. Both of these concepts fundamentally argue that the peripheral tissues would be, in this state, hypothyroid.
4. A popular view is that NTI patients are hypothyroid, but that this is a "good" state that avoids the alleged catabolic effects of normal thyroid hormone levels and is a desirable metabolic response. There is, of course, no factual basis for this explanation, despite its current widespread acceptance.
5. A final interpretation is that the patients are hypothyroid, but the treatment is useless or possibly harmful.

While admitting that I, as others, do not understand the true physiology of the problem, I believe that many of the arguments are flawed.

1. As indicated above, the true free thyroid hormone value cannot be determined by any current method in a manner that can be considered reliable.
2. While an abnormality of binding of hormone to proteins in blood may well be present, it is not obvious how important this factor is. Administration of replacement doses of L-thyroxine (such as 1.5 μg of thyroxine per kilogram/day i.v.) uniformly returns thyroxine and FTI values in NTI patients to normal, as shown by Brent et al[52]. Thus the binding abnormality cannot be the only, or even the major, aspect of the problem, and another explanation for the altered hormone levels must be considered.
3. To state that this situation is a "good" metabolic state is, in my view, totally unfounded. Can one similarly construe the fever of 105°F in infection, the blood sugar of 600 in diabetes, the low hematocrit of severe iron deficiency, and so forth, to represent the "wisdom of the body"? Usually normal T_4 and T_3 levels are considered beneficial for normal anabolic function. If it is conceived that a diminished hormone supply and therefore lessened anabolic function would be beneficial in disease, it would be logical by extension of this theory to treat all patients with a normal T_4, who are ill, with propylthiouracil to induce a lowering of thyroxine. No one is brave or foolish enough to make such a contention. Also, if one really believes that the free T_4 by dialysis,

or other assay system, can reliably indicate the free hormone level, and also that even normal levels of T_4 can be disadvantageous to an ill patient, then certainly patients who have elevated free T_4 levels, as detected by these tests, should be similarly treated with propylthiouracil to reduce levels to normal or low. Again, no one supports this contention, although it is an obvious part of the theory that a low hormone level is beneficial.

4. The possibility that therapy is disadvantageous is frequently presented. There are no data that support this view. Administration of thyroid hormone in low or replacement dosage to hypothyroid patients, in situations in which hypothyroidism is obvious, is always attempted, with due consideration for complicating factors such as cardiovascular dysfunction. Thyroid hormone therapy has been tried in two well-controlled and several less-controlled studies. The excellent study of Brent et al,[52] in which only thyroxine was given, and in which extremely ill patients were studied, showed neither benefit nor harm in a small group of patients so ill that 75% of each group died. A study of severely burned patients given 200 μg of T_3 daily also showed no beneficial or negative effect in a limited study.[38] In one uncontrolled study of neonates, replacement therapy with T_3 appeared to be helpful. In a study of animals with experimental infections, the administration of thyroid hormone appeared to accelerate the time of death, although final death rates were unchanged. The data are limited and so far prove neither harm nor benefit.
5. The tissue levels of both hormone and receptors are probably low in such patients, but little data are available. Probably, if therapy is to be carried out, it needs to be given as a mixture of triiodothyronine and thyroxine, since deiodination of thyroxine is clearly restricted in such patients. Further, it must be recognized that administration of replacement thyroid hormone is but one of many efforts made concurrently to buttress the clinical status of such severely ill patients, and one must also recognize that many other abnormalities are presumably present and as yet undetected.

Evidence for a low metabolic state in these patients includes the low measured free hormone levels in many assay systems, the reported low T_3 tissue levels, the fact that TSH is suppressed from elevated to normal levels in some hypothyroid patients during nonthyroidal illness, the lack of elevation of reverse T_3 in many NTI patients, and the recovery of TSH values to an elevated level following recovery from NTI.

Thus is seems appropriate to argue that there is no certainty about the actual tissue T_3 receptor occupancy in such patients but much evidence to suggest that it is low. There is no evidence that replacement therapy should be or has been proven to be disadvantageous. There are no data to suggest that replacement doses of thyroid hormone given to normal or hypothyroid patients would be harmful. Treatment of even subclinical hypothyroidism is reported to be valuable.[53] There is every reason to try all possible supportive measures for patients who are so critically ill and who have such a serious prognosis. While recognizing that much more information is needed, it seems prudent, on the basis of our knowledge up to this time, to give partial replacement therapy routinely in patients with nonthyroidal illness by administering 75 to 100 μg of thyroxine IV and 25 μg triiodothyronine orally, if possible.

Response by Dr. Leonard Wartofsky

Dr. DeGroot's response still leaves me with doubts. He questions the reliability of free T_4 assays in nonthyroidal illness (NTI), but later suggests that low values might indicate a possible benefit from T_4 therapy. However, the recent demonstration of normal free T_4 by immunochemiluminometric assay[54] may indicate that low T_4 values in NTI are only methodologic errors.

I agree that it would be foolish to construe that fever, hyperglycemia, or anemia reflect "body wisdom." However, the thyroid response to these perturbations may be an attempt to maintain homeostasis by avoiding the additional catabolism afforded by activation of T_4 to T_3.

Dr. DeGroot asserts that there are no data to support the view that thyroid hormone therapy is disadvantageous. However, one report on rats suggested an accelerated, although not increased, mortality from thyroid hormone after experimental infection.[55] Also, Brent and Hershman[39] showed a possibly accelerated, although not increased, mortality rate in T_4-treated critically ill patients.

Dr. DeGroot correctly states that there is little data from humans to indicate that hormone or hormone receptor levels in the tissues are low. However, it has been repeatedly shown in animals that reduced caloric intake will reduce T_3 receptors,[56,57] as will uremia[58,59] and other models of NTI.[60] Liver content was also low in liver disease. However, can one reliably conlcude that the tissues are "hypothyroid" because T_3 receptor binding is reduced? Dr. DeGroot himself[61] has described marked reductions in liver nuclear T_3 content in starvation but *not* in hypothyroidism and concluded that "a tenable hypothesis is that the changes represent a group of selective physiologic responses designed to . . . protect tissues from the catabolic effects of T_3 and T_4. . . ."

Dr. DeGroot advises a mixture of T_4 and T_3 because T_4 to T_3 conversion is reduced, and the active hormone, T_3, is needed. Why then give T_4 at all? That access of T_4 to tissue sites is not impaired is apparent from the continuing 5-deiodination (reverse T_3 production) seen in NTI. That reduced T_3 production serves to minimize further protein (and tissue) breakdown is apparent from the major increases in fat catabolism[62] and protein catabolism seen with administration of small doses of T_3.[63,64]

I would urge continuing skepticism on this point until additional studies elucidate the uncertainties underlying the above questions, following the principle of "primum non nocere."

Response by Dr. Leslie DeGroot

We recommend full nutritional support to avoid starvation because that is a far more important cause of tissue catabolism than a T_3 supplement. We use T_4 because these patients are usually maintained by intravenous fluids, and there is no conveniently available preparation of T_3 for intravenous use. The reason for this difference in approach is that we do not know whether treatment or no treatment is the "nocere." At least in our hands, the recommended treatment has not seemed to be harmful.

References

1. Charbonneau M, Dussault JH, Refetoff S, et al.: A study of the inheritance of thyroxine-binding globulin (TBG) deficiency from data obtained in 13 families detected by a neonatal screening program. Clin Invest Med, (submitted).
2. Moses AC, Lawlor J, Haddow J, Jackson IMD: Familial euthyroid hyperthyroxinemia resulting from increased thyroxine binding to thyroxine-binding prealbumin. N Engl J Med 1982; 306:966.
3. Rajatanavin R, Fournier L, DeCosimo D, et al: Elevated serum free thyroxine by thyroxine analog radioimmunoassays in euthyroid patients with familial dysalbuminemic hyperthyroxinemia. Ann Intern Med 1982;97:865.
4. Ruiz M, Rajatanavin R, Young RA, et al: Familial dysalbuminemic hyperthyroxinemia. N Engl J Med 1982; 306:635.
5. Rajatanavin R, and Braverman LE: Euthyroid hyperthyroxinemia. J Endocrinol Invest 1983;6:493.
6. Chan V, Wang C, Yeung RTT: Effects of heroin addiction on thyrotrophin, thyroid hormones, and prolactin secretion in men. Clin Endocrinol 1979;10:557.
7. Sakara S. Nakamura S, Miura K: Autoantibodies against thyroid hormones or iodothyronine. Ann Intern Med 1985;103:579.
8. Stockigt JR, Dyer SA, Mohr VS., et al.: Specific methods to identify plasma binding abnormalities in euthyroid hyperthyroxinemia. J Clin Endocrinol Metab 1986;62:230.
9. Larsen PR, Silva JE, Kaplan MM: Relationships between circulating and intracellular thyroid hormones: Physiological and clinical implications. Endocr Rev 1981;2:87.
10. Birkhauser M, Busset R, Burer T, Burger A: Diagnosis of hyperthyroidism when serum-thyroxine alone is raised. Lancel 1977;2:53.
11. Bober SA, McGill AC, Tunbridge WMG: Thyroid function in hyperemesis gravidarum. Acta Endocrinol 1986;111:404.
12. Galton VA: Some effects of altitude on thyroid function. Endocrinology 1972;91:1393.
13. Wenzel KW: Pharmacological interference with in vitro tests of thyroid function. Metabolism 1981;30:717.

14. Burger A, Dinichert D, Nicod P, et al: Effect of amiodarone on serum triiodothyronine, reverse triiodothyronine, thyroxine, and thyrotropin. J Clin Invest 1976;58:255.
15. Martino E, Safran M, Aghini-Lombardi F, et al.: Environmental iodine intake and thyroid dysfunction during chronic amiodarone therapy. Ann Intern Med 1984;101:28.
16. Gow SM, Caldwell G, Toft AD, et al: Relationship between pituitary and other target organ responsiveness in hypothyroid patients receiving thyroxine replacement. J Clin Endocrinol Metab 1987;64:364.
17. Brown ME, Refetoff S: Transient elevation of serum thyroid hormone concentration after initiation of replacement therapy in myxedema. Ann Intern Med 1980;92:491.
18. Ingbar JC, Borges M, Iflah S, et al.: Elevated serum thyroxine concentration in patients receiving "replacement" doses of levothyroxine. J Endocrinol Invest 1982;5:77.
19. Refetoff S, DeWind LT, DeGroot LJ: Familial syndrome combining deafmutism, stippled epiphyses, goiter and abnormally high PBI: Possible target organ refractoriness to thyroid hormone. J. Clin Endocrinol Metab 1967;27:279.
20. Sarne DH, Refetoff S, Rosenfield RL, Farriaux JP: Sex hormone binding globulin in the diagnosis of peripheral tissue resistance to thyroid hormone: The value of changes following short-term triiodothyronine administration. J Clin Endocrinol Metab, in press.
21. Weintraub BD, Gershengorn MC, Kourides IA, Fein H: Inappropriate secretion of thyroid stimulating hormone. Ann Intern Med 1981; 95:339.
22. Sarne DH, Refetoff S, Murata Y, et al: Variant thyroxine-binding globulin in serum of Australian Aborigines. A comparison with familial TBG deficiency in Caucasians and American Blacks. J Endocrinol Invest 1985;8:217.
23. Oppenheimer JH, Werner SC: Effect of prednisone on thyroxine binding proteins. J Clin Endocrinol Metab 1966;26:715.
24. Larsen PR: Salicylate-induced increases in free triiodothyronine in human serum: Evidence of inhibition of triiodothyronine binding to thyroxine-binding globulin and thyroxine-binding prealbumin. J. Clin Invest 1972;51:1125.
25. Suzuki H, Yamazaki N, Suzuki Y, et al: Lowering effect of diphenylhydantoin on serum free thyroxine and thyoxine binding globulin (TBG). Acta Endocrinol 1984;105:477.
26. Lazarus JH, John R, Bennie EH, et al: Lithium therapy and thyroid function: A long-term study. Psychol Med 1981;11:85.
27. Chopra IJ, Solomon DH, Hepner GW, Morgenstein AA: Misleadingly low free thyroxine index and usefulness of reverse triiodothyronine measurement in nonthyroidal illnesses. Ann Intern Med 1979;90:905.
28. Wartofsky L, Burman KD: Alterations in thyroid function in sick patients with systemic illness: The "euthyroid sick syndrome". Endocr Rev 1982;3:164.
29. Chopra IJ, Hershman JM, Partridge WM, Nicoloff JT: Thyroid function in nonthyroidal illness. Ann Intern Med 1983;98:946.
30. Chopra IJ, Juang T-S, Hurd RE, et al: A competitive ligand binding assay for measurement of thyroid hormone-binding inhibitor in serum and tissues. J Clin Endocrinol Metab 1984;58:619.
31. Huang TS, Hurd RE, Chopra IJ, et al: Coexistence of inhibitors of thyroid hormone binding and phagocytosis in extrathyroidal tissues. Abstract pre-

sented at the 58th Annual Meeting of the American Thyroid Association, September 22–25, 1982, p T-28 in Book of Abstracts.
32. Mendel CM, Frost PH, Cavalieri RR: Effect of free fatty acids on the concentration of free thyroxine in human serum: The role of albumin. J Clin Endocrinol Metab 1986;63:1294.
33. Reichlin S, Bolinger J, Nejad I, Sullivan P: Tissue thyroid hormone concentration of rat and man determined by radioimmunoassay: Biologic significance. Mt Sinai J Med 1973;40:502.
34. Nicoloff JT: Thyroid function in nonthyroidal disease. Chapter In: DeGroot LJ, ed. *Endocrinology*, 2nd ed. Philadelphia: WB Saunders, in press.
35. Whehmann RE, Gregerman RI, Burns WH, et al: Suppression of thyrotropin in the low-thyroxine state of severe nonthyroidal illness. N Engl J Med 1985;312:546.
36. Hamblin PS, Dyer SA, Mohr VS, et al: Relationship between thyrotropin and thyroxine changes during recovery from severe hypothyroxinemia of critical illness. J Clin Endocrinol Metab 1986;62:717.
37. Hesch RD, Husch M, Kodding R, et al: Treatment of dopamine-dependent shock with triiodothyronine. Endocr Res Commun 1981;8:229.
38. Becker RA, Vaughan GM, Ziegler MG, et al: Hypermetabolic low triiodothyronine syndrome of burn injury. Crit Care Med 1982;10:870.
39. Brent GA, Hershman JM: Thyroxine therapy in patients with severe nonthyroidal illnesses and low serum thyroxine concentration. J Clin Endocrinol Metab 1986;63:1.
40. Saberi M, Utiger RD: Serum thyroid hormone and thyrotropin concentrations during thyroxine and triiodothyronine therapy. J Clin Endocrinol Metab 1974;39:923.
41. Symons RG, Murphy LJ: Acute changes in thyroid function tests following ingestion of thyroxine. Clin Endocrinol 1983;19:539.
42. Wennlund A: Variation in serum levels of T_3, T_4, FT_4 and TSH during thyroxine replacement therapy. Acta Endocrinol (Copenh) 1986;113:47.
43. Melmed S, Geola FL, Reed AW, et al: A comparison of methods for assessing thyroid function in nonthyroidal illness. J Clin Endocrinol Metab 1982;54:300.
44. Chopra IJ, Van Herle AJ, Chua Teco GN, Nguyen AH: Serum free thyroxine in thyroidal and nonthyroidal illnesses: A comparison of measurements by radioimmunoassay equilibrium dialysis, and free thyroxine index. J Clin Endocrinol Metab 1980;51:135.
45. Kaplan MM, Larsen PR, Crantz FR, et al: Prevalence of abnormal thyroid function test results in patients with acute medical illnesses. Am J Med 1982;72:9.
46. Wood DG, Cyrus J, Samols E: Low T_4 and low FT_4I in seriously ill patients: Concise communication. J Nucl Med 1980;21:432.
47. Kaptein EM, Levitan D, Feinstein EI, et al: Alterations of thyroid hormone indices in acute renal failure and in acute critical illness with and without acute renal failure. Am J Nephrol 1981;1:138.
48. Kaptein EM, MacIntyre SS, Weiner JM, et al: Free thyroxine estimates in nonthyroidal illness: Comparison of eight methods. J Clinic Endocrinol Metab 1981;52:1073.
49. Slag MF, Morley JE, Elson MK, et al: Free thyroxine levels in critically ill patients. A comparison of currently available assays. JAMA 1981;246:2702.

50. Wang Y-S, Pekary AE, England ML, Hershman JM: Comparison of a new ultrafiltration method for serum free T_4 and free T_3 with two RIA kits in eight groups of patients. J Endocrinol Invest 1985;8:495.
51. Braverman LE, Abreau CM, Brock P, et al: Measurement of serum free thyroxine by RIA in various clinical states. J Nucl Med 1980;21:233.
52. Brent GA, Hershman JM, Reed AW, et al: Serum angiotensin-tensin-converting enzyme in severe non-thyroidal illnesses associated with low serum thyroxine concentrations. Ann Intern Med 1984;100:680.
53. Cooper DS, Halpern R, Wood LC, et al: L-thyroxine therapy in subclinical hypothyroidism, a double blind, placebo-controlled trial. Ann Intern Med 1984;101:18.
54. Sturgess ML, Weeks I, Evans PJ, et al: An immunochemiluminometric assay for serum free thyroxine. Clin Endocrinol 1987;27:383.
55. Little JS: Effect of thyroid hormone supplementation on survival after bacterial infection. Endocrinology 1985;117:1431.
56. Burman KD, Lukes Y, Wright FD, Wartofsky L: Reduction in hepatic triiodothyronnine binding capacity induced by fasting. Endocrinology 1977;101:1331.
57. Schussler GC: Fasting decreases triiodothyronine receptor capacity. Science 1978;199:686.
58. Thompson P Jr, Burman KD, Lukes YG, et al: Uremia decreases nuclear 3,5,3′-triiodothyronine receptors in rats. Endocrinology 1980;107:1081.
59. Lim VS, Passo C, Murata Y, et al: Reduced triiodothyronine content in liver but not pituitary of the uremic rat model: Demonstration of changes compatible with thyroid hormone deficiency in liver only. Endocrinology 1984;114:280.
60. Surks ML, Grajower MM, Tai M, DeFesi CR: Decreased hepatic nuclear L-triiodothyronine receptors in rats and mice bearing transplantable neoplasms. Endocrinology 1978;103:2234.
61. DeGroot LJ, Coleoni AH, Rue PA, et al.: Reduced nuclear triiodothyronine receptors in starvation-induced hypothyroidism. Biochem Biophy Res Commun 1977;79:173.
62. Carter WJ, Shakir KM, Hodges S, et al: Effect of thyroid hormone on metabolic adaptation to fasting. Metabolism 1974;24:1177.
63. Lim VS, Flanigan MF, Zavela DC, Freeman RM: Protective adaptation of low serum triiodothyronine in patients with chronic renal failure. Kidney Int 1985;28:541.
64. Gardner DF, Kaplan MM, Stanley CA, Utiger RD: Effect of triiodothyronine replacement on the metabolic and pituitary responses to starvation. N Engl J Med 1979;300:579.

2

Clinical Importance of Abnormal Thyroid Function Test Results in Patients with Nonthyroidal Illness

LEONARD WARTOFSKY

The multiple effects of systemic illness on thyroidal economy have been recently reviewed,[1-3] and are commonly referred to as the "euthyroid sick syndrome." Implicit in this designation is that such patients are indeed euthyroid, but manifest a number of "abnormal" thyroid function tests. One may gain some insight into the widespread and complex aberrations seen in severe illness with the appreciation that no single thyroid function test is diagnostic or without pitfalls in such patients. The hallmark laboratory parameters of the euthyroid sick syndrome include a normal to markedly decreased serum total T_4 (usually in proportion to the severity of the illness), and a low to undetectable serum total T_3. However, the serum total T_4 and free T_4 index values may be elevated in the absence of thyroid disease ("euthyroid hyperthyroxinemia"), and hyperthyroidism must be ruled out in such cases.

Low T_3 Syndrome

Since many (if not all) of the clinical states that are associated with the euthyroid sick syndrome are marked by a decreased total and free T_3, these patients also have been described as having the "low T_3 syndrome". The low serum T_3 is due to reduced activity of the 5′-deiodinase, which activates the conversion of T_4 to T_3 by 5-monodeiodination. Instead, the degradation of T_4 in the sick patient proceeds by an alternative (inactivating) pathway; this pathway is subserved by 5′-deiodinase and results in formation of reverse T_3. Reverse T_3 is calorigenically inactive relative to T_4 or T_3, and this "switch" in deiodination in systemic illness has been interpreted teleologically as an homeostatic mechanism for energy conservation. Thus, concomitant with the fall in serum T_4 and T_3 usually seen with systemic illness, there is a rise in values for reverse T_3. This is due largely to reduced monodeiodination and clearance of reverse T_3, which is itself metabolized via the same 5′-deiodinase enzyme that converts T_4 to T_3.

In general, the aberrations in thyroid function tests accompanying systemic illness (low T_4, low T_3, high reverse T_3) are correlated inversely to the severity of the nonthyroidal illness. Indeed, there is even a clear correlation with prognosis, the most marked alterations in thyroid function tests being associated with poor recovery or death.[4] In turn, clinical recovery and the associated improvement in nutrition will be marked by a return of serum T_4, T_3, and reverse T_3 to normal levels, and in some cases to transient increases in T_4 or free T_4 index to supranormal levels.

In spite of its limited utility, many hospitals still routinely provide measurement of the resin T_3 uptake, and the results of this test in sick patients are likely to add to the clinician's confusion. Higher values for resin T_3 uptake are obtained when thyroid hormone-binding protein affinity or capacity is reduced, and this is the case in systemic illness that is marked by the presence of circulating inhibitors to hormone binding.[5,6] The free T_4 index, as a product of the total T_4 and resin T_3 uptake values, may be a valid index of euthyroidism (correcting for the high resin T_3 uptake with the low T_4 level), but this tends to be true only in mild to moderately severe systemic illness, and even the free T_4 index will be misleading in critically ill patients.[7]

Measurement of serum thyroid-stimulating hormone (TSH) is generally believed to be the most sensitive discriminant between normal and failing thyroid gland function. Most patients with abnormally low serum total T_4 and T_3 levels accompanying systemic illness have values for TSH within the normal range,[2,8] a finding that has constituted perhaps the strongest evidence that these patients are indeed euthyroid. However, some sick patients have low serum TSH levels,[9-12] a phenomenon that has become better appreciated with the recent advent of more sensitive assay techniques.[13] When TSH levels fail to rise in the face of low levels of free T_4 and free T_3, suggested explanations have included the presence of a circulating inhibitor acting on the pituitary (e.g., stress, corticosteroids, fever, drugs, etc.) or the release of biologically inactive TSH. Evidence for the latter phenomenon has come from the studies of Lee et al[14] demonstrating that the TSH secreted in sick patients may have altered glycosylation and reduced biologic activity. This phenomenon could explain apparently paradoxically reduced thyroidal secretion in the face of normal TSH levels by radioimmunoassay (RIA), as well as the slight elevations in radioimmunoassayable TSH sometimes seen in sick patients such as those with cirrhosis.[15] Recent availability of ultrasensitive assay techniques for TSH has allowed some further insight into pituitary TSH secretion during illness. Spencer[16] has observed a wide range of values from levels in the suppressed range (0.1–0.4 μU/ml) to very clearly normal levels between 0.5 and 5.0 μU/ml; there is a tendency for those patients with the very low or suppressed TSH concentrations to have had some other factor operative, such as dopamine infusion, which could account for the very low levels.

Hyperthyroxinemia in Acute Illness

There are a few examples of nonthyroidal illness that are usually accompanied by increases rather than decreases in serum total T_4 and total T_3. In such patients, the acute illness may be associated with fever, tremors, rapid heart rate, and anxiety, which taken together with an elevated result for total T_3 or total T_4 could lead to an erroneous diagnosis of hyperthyroidism. In these patients, a free T_4 or free T_3 determination is indicated to help rule out this complication and thereby avoid unnecessary therapy.

Liver Disease

In acute infectious hepatitis, there is release from the liver of increased amounts of α-globulin resulting in an augmented thyroxine-binding globulin (TBG) binding capacity. The consequent increase in serum total T_4 and total T_3 due to increased binding is accompanied by a reduced resin T_3 uptake and the free T_4 index or free T_3 index is normal, as is the serum TSH, confirming the euthyroid status of these patients. In alcoholic hepatitis or liver cirrhosis, the free T_4 index, T_4/TBG, and free T_3 index are all decreased while the free T_4 or free T_3 by RIA are again normal.[17] The situation is a little more complicated in chronic active hepatitis and primary biliary cirrhosis, in which increases in total T_4 and total T_4 with reduced resin T_3 uptake again suggest increases in TBG.[18] While Liewendahl et al[17] have reported normal free T_4 and T_3 (with reduced free T_3 index) in these patients, a large number of them will have decreased free T_4 and free T_3 (in spite of an increased percentage of free hormone). This latter phenomenon may be on the basis of coincident underlying Hashimoto's thyroiditis with early thyroid failure.[18]

Critical Illness

Generally speaking, the more severe the systemic illness, the more profound will be the altered thyroid function tests comprising the euthyroid sick syndrome (e.g., the low total T_4 and total T_3). There is a population of sick patients, however, who manifest increases (rather than decreases) of total T_4 into the thyrotoxic range, usually with normal total T_3. In a series of 31 such patients reported by Birkhauser et al,[19] half of them ultimately manifested a rise in total T_3 consistent with hyperthyroidism. In a smaller series of 13 patients reported by Gavin et al,[20] three subjects subsequently developed hyperthyroidism. While some workers have cautioned against the reliability of the free T_4 measurement in such patients,[21] I believe that the differential diagnosis in regard to ruling in or out hyperthyroidism rests with the free T_4 and the response to thyrotropin-releasing hormone (TRH). This belief is based on the presence of an initially normal, rather than decreased serum total T_3 as is usually seen in such sick patients, and the fact that so many of them do ultimately manifest more overt and classic hyperthyroidism. Moreover, it is known that iodine excess will precipitate hyperthyroidism in subjects with normal thyroid glands as well as in those with underlying autonomous function (euthyroid Graves disease; solitary autonomous nodules), and sick patients are often recipients of massive iodine loads in the form of X-ray contrast media. Consistent with this speculation is the fact that iodine-induced hyperthyroidism is often marked initially by excess thyroxine production relative to triiodothyronine ("T_4-toxicosis") as was seen in many of the patients previously reported.[19,20] Since iodine contamination diminishes the utility of the 24-hour radioiodine uptake test, and the free T_3 may be reduced due to the complicating systemic illness, the best clues to the true diagnosis remain the free T_4 and the TRH response, which, if normal, exclude thyrotoxicosis.

Acute Psychiatric Disease

Since both thyrotoxicosis and myxedema may be accompanied by psychiatric manifestations, it is common practice to draw blood samples for thyroid function studies during the routine evaluation of mental disorders or when patients are admitted to general psychiatric wards. This practice has uncovered a large fraction of the psychiatric population with increased serum T_4 of uncertain etiology. As reviewed previously,[2] both total T_4 and free T_4 index are increased acutely, but then normalize after specific therapy for the psychiatric illness. In acute depression, serum free T_4 has been normal to slightly increased while free T_3 is normal. The TRH response is of less help than usual in these disorders because it may be blunted without evidence of excess circulating thyroid hormone concentrations. An elevated free T_4 index has been seen in 7% to 9%[22–24] of general psychiatric admissions, and an elevated free T_3 index has been seen in 14%.[23] Understanding of the underlying abnormality is complicated by the fact that many of these patients are taking drugs, some of which (amphetamines) are known to elevate serum thyroxine. The data presently available do not allow a determination of whether or not this represents true, albeit transient, thyrotoxicosis. This perhaps could be clarified in future studies of such patients, if the responsible investigators obtain concomitant parameters of metabolic status (basal metabolic rate; pulse wave arrival time (QkD); Achilles tendon half-relaxation time) as well as estimates of free T_4 by either equilibrium dialysis or a comparably accurate direct RIA technique. Until those data are available, serial determinations of total and free T_4 are warranted on a weekly basis; the patients with transient hyperthyroxinemia should declare themselves within 2 to 3 weeks. Alternatively, a free T_3 could be performed, and if normal would rule out any complicating hyperthyroidism with the exception of "T_4 toxicosis." This variety of the disease is rare, and would be excluded by a normal TSH response to TRH (if obtainable).

Given a low serum T_4 and T_3, the critical question facing the physician is whether the patient is hypothyroid or not, since hypothyroidism accompanying systemic illness would have an adverse effect on recovery and survival. On the other hand, the clinician cannot injudiciously begin thyroid hormone therapy in such patients without being confident of the diagnosis, since such treatment might provoke potentially fatal events such as myocardial infarction or dangerous arrhythmias. Indeed, thyroid hormone supplements to sick patients or experimental animals have not improved, and rather may have worsened, survival.[25,26] Thus, the physician must discriminate between the two clinical situations of systemic illness with hypothyroidism versus merely aberrated thyroid function tests representing a homeostatic mechanism to reduce caloric expenditure in a sick but euthyroid individual.

Most workers in the field have attempted to resolve the clinical dilemma and answer the latter question by documenting the presence or absence of euthyroidism with the measurement of either free hormone levels, serum TSH, or the TSH response to TRH. Since TSH and TRH responses may be suppressed by a variety of factors present in sick individuals (fever, stress, high endogenous steroids, caloric deprivation, certain drugs), these tests have not proved reliable unless elevated results consistent with primary hypothyroidism are obtained. As a consequence, other workers have examined whether estimates of free T_4 concentration would serve as a valid index of true thyrometabolic status in patients with systemic illness. Use of the free T_4 index as the product of total T_4 and resin T_3 uptake produced values in sick patients that underestimated free T_4 concentration as compared to measurements by equilibrium dialysis.[27] A modified free T_4 index, expressing the index as the product of total T_4 and the ratio of resin ^{125}I-T_3, was found to be more reliable in some early reports,[27,28] but Kaptein et al[29] found this modified free T_4 index to be reliable primarily in the sick patients with normal total T_4, but not in those with low total T_4. This latter difficulty in distinguishing between hypothyroid sick and euthyroid sick individuals with the free T_4 index (regardless of how it is calculated) was also found by Melmed et al.[26] Throughout the literature, the "gold standard" for free T_4 levels was that value obtained by the tedious and expensive equilibrium dialysis technique, and the development of a rapid, accurate, and direct RIA measurement of free T_4 was a challenge and unattainable goal until just a few years ago.

There are now commercially available about two dozen rapid, inexpensive, and direct RIA systems for determination of free T_4. Several of the more popular ones have been critically evaluated in large numbers of both hypothyroid and euthyroid individuals with complicating systemic illnesses.[26,29,30] In the present context, the optimal assay would be one that could distinguish between hypothyroidism and the euthyroid sick syndrome. In several studies, the Clinical Assays Gammacoat system seemed to achieve this goal with the greatest reliability,[26,29] although Wood et al[31] found the Gammacoat system to accurately detect hypothyroidism in patients without other illness, but to give falsely normal values in sick hypothyroid patients. Using either this system or equilibrium dialysis, the serum free T_4 and free T_3 tend to be slightly reduced in patients with chronic renal failure, chronic active hepatitis, primary biliary cirrhosis, and after major surgical procedures. The free T_4 is normal to slightly increased in patients with nephrotic syndrome, starvation, diabetic ketoacidosis, and Laennec's cirrhosis. Free T_3, on the other hand, is reduced in most malnutritional states, and in liver diseases of both the acute and chronic types. Although this may be confusing and appear to devalue the merits of free hormone assessment in systemic illness, it must be remembered that the desirable assay need only exclude hypothyroidism, and either a normal or high free T_4 will serve this purpose. Until a fully reliable free T_4 assay in sick patients is available, the clinician will need to rely on his or her clinical impression of these patients and ancillary bits of data such as the serum TSH and reverse T_3. An occasional patient will still elude definitive diagnosis in spite of the sophisticated armamentarium of available tests. For that patient, the physician must fall back upon his or her clinical judgment, and also rely on time to help elucidate the true situation.

Discussion by Dr. John E. Freitas

Our experience with transient hyperthyroxinemia in acute psychiatric admissions parallels that of Morley and Shafer with schizophrenics and amphetamine abusers exhibiting transient hyperthyroxinemia most commonly.

Discussion by Dr. Joel I. Hamburger

My understanding is that with the new supersensitive TSH assays, even though the basal TSH level may be subnormal, it is seldom, if ever, undetectable in the euthyroid sick patient; and there is always more response to TRH than would be seen in hyperthyroidism. Is your experience in agreement with that concept?

You said that hypothyroidism would have an adverse effect on recovery and survival in the euthyroid sick patient. I am sure that is so for those with overt hypothyroidism. But those patients should not be difficult to diagnose because they will usually have goiter, positive antithyroid antibodies, a surgical scar, or a history of ^{131}I therapy, and also easily identifiable physical findings of hypothyroidism. Isn't the problem really that of diagnosing mild or even questionable hypothyroidism, arising perhaps on a central basis? Recognizing that there might be some uncertainty, would your best judgment be that failure to treat such patients with thyroid hormone would compromise their recovery? In answering this question, please take into consideration the lag time before the administration of thyroid hormone produces an effect.

Response by Dr. Leonard Wartofsky

You are correct that the basal TSH is usually measurable in sick patients when measured by an ultrasensitive TSH assay. Only approximately 1% of such patients will have an undetectable value[10,32] and in some series, mean TSH levels in fact may be slightly higher in sick patients than in euthyroid controls.[8,13] However, one recent report[33] showed that 13 of 34 critically ill patients had TSH levels <0.4 mU/L, but even these patients had reasonably good responses to TRH as a group. (Note that the latter workers employed a relatively less sensitive assay, thereby accounting for at least some of the "undetectable" levels.)

I do not believe that there are any convincing data that mild hypothyroidism has an adverse effect on recovery or survival of patients with systemic or critical illness, but the possibility is difficult to deny. You are again correct in the implication that the diagnosis of primary hypothyroidism in the sick patient should pose little problem in the presence of low T_4, T_3, free T_4, and an elevated TSH, and rather, the difficulty pertains to ruling out hypothyroidism secondary to hypothalamic or pituitary disease in the patient with low T_4, T_3, free T_4, *and* TSH with a marginal or blunted response to TRH. Often, such patients may have a history of receiving agents that could suppress the pituitary (steroids, dopamine, etc). TSH responses to TRH may be blunted in patients with central nervous system disorders[34] and we should also keep in mind the rare occurrence of suprahypophyseal hypothyroidism that may ensue following prolonged coma after head trauma.[35]

Discussion by Dr. Ian D. Hay

In the situation of hyperthyroxinemia associated with critical illness, Dr. Wartofsky would seek assistance from measuring the serum free T_4 and the TRH response. However, when faced with the problem of a sick patient with a low serum T_4 and T_3, he favors only measuring a free T_4 level, since TSH and TRH responses have not proved reliable unless elevated results are obtained. These two viewpoints seem to me somewhat contradictory.

I certainly agree that no currently available free T_4 assay seems "fully reliable" in sick patients. I wonder, however, what significance (if any) Dr. Wartofsky would place on the finding of a normal (euthyroid) serum TSH level (as measured in a sensitive immunometric assay) in the context of nonthyroidal illness. Could this finding provide comparable information to a "normal or high free T_4" as measured by the Gammacoat system or by equilibrium dialysis? Does Dr. Wartofsky really believe that there is no role for supersensitive TSH (STSH) measurement in the hospitalized patient with nonthyroidal illness?

Response by Dr. Leonard Wartofsky

The intent of the statement quoted by Dr. Hay was to point out to practitioners that the diagnosis of primary hypothyroidism in a sick patient may be rendered more difficult because of a number of factors that could alter the expected high basal TSH and exaggerated TSH response to TRH. These include:

1. Suppression of TSH and TSH response to TRH by starvation and malnutrition[36,37] or on the contrary;
2. Mild increases in basal TSH in sick patients[15,38] or transient increases in TSH during recovery from critical illness;[10,12] and
3. Alterations of TSH or TSH response to TRH by drugs (e.g., steroids, iodinated contrast agents, amiodarone, dopamine, etc.).

These interferences would be trumped by a clear-cut elevation in basal TSH or TSH response to TRH strongly suggesting primary hypothyroidism. My final paragraphs were intended to place in historical perspective the use of free T_4 assays by several workers in an attempt to clarify this dilemma, but I was not advocating the utility of a free T_4 measurement as necessarily superior to a TSH measurement. At the practical level in a given confusing or subtle clinical problem, accurate differentiation often will require values for both the free T_4 and TSH.

Again, I did not mean to imply that there was no role for an ultrasensitive TSH measurement in the hospitalized patient with nonthyroidal illness. I believe that available data suggest that STSH should be "normal" in sick patients,[10,32] but may be occasionally slightly higher than normal[8] or slightly lower (undetectable), particularly with the less sensitive of the ultrasensitive assays.[33] Therefore, I would place moderately strong confidence in normal TSH values (as indicated by Dr. Hay) in this situation. However, it remains to be seen whether or not any current free T_4 assay will prove "fully reliable" in sick patients. The recently reported immunochemiluminometric assay for free T_4 is very impressive in this regard,[39] clearly differentiating most euthyroid sick and hypothyroid patients.

Discussion by Dr. James C. Sisson

In patients with critical illness who are suspected of having concurrent hyperthyroidism, Dr. Wartofsky recommends a TRH test when other data leave uncertainty of diagnosis. Will not the sensitive TSH assay concentration be as good as the results of the TRH test? Will the factors that affect the TSH levels in critical illness also affect the TSH response to TRH?

The same question will apply to the use of the TRH test in patients with acute psychiatric illness. If a normal response of TSH to TRH will exclude "T_4-toxicosis," will not a normal basal sensitive TSH level permit the same conclusion?

When the question of hypothyroidism is raised in a patient with severe nonthyroidal illness, it would seem that an elevated TSH concentration, particularly a level over 30 μU/ml, would indicate the need for treatment of primary hypothyroidism. However, if pituitary hypothyroidism is a reasonable possibility, the use of a reverse T_3 assay is said to help distinguish between hypothyroidism and the effects of the nonthyroidal illness on the usual tests of thyroid function. Are the results of the reverse T_3 assay, in fact, discriminating in this circumstance? The implications of the result extend beyond that of thyroid hormone treatment and would include concern for the level of cortisol secretion and the nature of the pituitary disease.

Response by Dr. Leonard Wartofsky

Dr. Sisson is correct in his implication that the ultrasensitive TSH measurement should prove as valid or useful as a TRH test in detecting the patient with hyperthyroidism complicating critical illness. This has been our experience and it is supported by data from others.[8,33,40]

The situation is a bit more complex in psychiatric patients who may experience hyperthyroxinemia and blunted TRH responses on presentation,[41,42] although this is not necessarily the case. Although there are little published data to date on ultrasensitive TSH results in psychiatric patients, I can think of no reason why basal TSH measured by these methods should not provide as much information as TRH testing.

I am not sure that I understand the specific thrust of the question about "pituitary" hypothyroidism and systemic illness. One might see low T_4, T_3, free T_4, TSH, and TSH response to TRH. In contrast, the "euthyroid sick" patient might display low T_4 and T_3 but a normal basal STSH and a normal to slightly blunted TRH response. The latter patient (except perhaps one with renal disease) should also demonstrate a high serum reverse T_3 level, whereas hypothyroid patients (primary or pituitary) should have low reverse T_3 levels. I believe that Dr. Sisson's question relates to the problem of the sick hypothyroid patient in regard to whether the anticipated high reverse T_3 of sickness or the low reverse T_3 of hypothyroidism predominates. Wood[43] indicated that the reverse T_3 was not a valid criterion for hypothyroidism because it still could be increased in those sick patients with mild hypothyroidism and only marginally low free T_3; those sick patients with the lower free T_4 and T_3 levels did have a low reverse T_3 as well. Dr. Sisson is correct in questioning the potential practical value of marginal results in a specific patient suspected of having hypothyroidism complicating systemic illness, although low reverse T_3

levels would support the diagnosis of hypothyroidism. Perhaps the resolution of the dilemma will be aided by the use of the newest immunochemiluminometric assay for free T_4,[39] which appears to differentiate clearly between hypothyroid and sick patients.

References

1. Chopra IJ, Hershman JM, Pardridge WM et al: Thyroid function in nonthyroidal illnesses. Ann Intern Med 1983;98:946.
2. Wartofsky L, Burman K: Alterations in thyroid function in patients with systemic illnesses: The euthyroid sick syndrome. Endocr Rev 1982;3:164.
3. Brent GA, Hershman JM: Effects of nonthyroidal illness on thyroid function tests. In: VanMiddlesworth L, ed. The Thyroid Gland. Chicago: Year Book, 1986, p 83.
4. Slag MF, Morley JE, Elson MK: Hypothyroxinemia in critically ill patients as a predictor of high mortality. JAMA 1981;245:43.
5. Chopra IJ, Chua Teco GN, Nguyen AH, Solomon DH: In search of an inhibitor of thyroid hormone binding to serum proteins in nonthyroid illnesses. J Clin Endocrinol Metab 1979;49:63.
6. Oppenheimer JH, Schwartz HL, Mariash CN, Kaiser FE: Evidence for a factor in the sera of patients with nonthyroidal disease which inhibits iodothyronine binding by solid matrices, serum proteins and rat hepatocytes. J Clin Endocrinol Metab 1982:54:757.
7. Chopra IJ, Solomon DH, Hepner GW, Morganstein AA: Misleadingly low free thyroxine index and usefulness of reverse triiodothyronine measurement in nonthyroidal illnesses. Ann Intern Med 1979;90:905.
8. Basset F, Eastman CJ, Ma G, et al: Diagnostic value of thyrotropin concentrations in serum as measured by a sensitive immunoradiometric assay. Clin Chem 1986;32:461.
9. Wehmann RE, Gregerman RI, Burns WH, et al: Suppression of thyrotrophin in the low-thyroxine state of severe nonthyroidal illness. N Engl J Med 1985;312:546.
10. Hamblin PS, Dyer SA, Mohr VS, et al: Relationship between thyrotropin and thyroxine changes during recovery from severe hypothyroxinemia of critical illness. J Clin Endocrinal Metab 1986;62:717.
11. Bacci V, Schussler GC, Kaplan TB: The relationship between serum triiodothyronine and thyrotrophin during systemic illness. J Clin Endocrinol Metab 1982;54:1229.
12. Brent GA, Hershman JM, Braunstein GD: Patients with severe nonthyroidal illness and serum thyrotrophin concentrations in the hypothyroid range. Am J Med 1986; 81:463.
13. Hardy MJ, Ragbeer SS, Nascimento L: Pituitary-thyroid function in chronic renal failure assessed by a highly sensitive thyrotropin assay. J Clin Endocrinal Metab 1988;66:233.
14. Lee H-Y, Suhl J, Pekary AE, Hershman JM: Secretion of thyrotropin with reduced Concanavalin-A-binding activity in patients with severe nonthyroid illness. J Clin Endocrinol Metab 1987;65:942.
15. Chopra IJ, Solomon SH, Chopra U, et al: Alterations in circulating thyroid hormones and thyrotropin in hepatic cirrhosis: Evidence for euthyroidism

despite subnormal serum triiodothyronine. J Clin Endocrinol Metab 1974;39:501.
16. Spencer C: Clinical utility of sensitive TSH assays. Thyroid Today 1986:IX (No. 2) (Flint Laboratories).
17. Liewendahl K, Helenius T, Tanner P, Salaspuro M: Serum free thyroid hormone concentrations and indices in alcoholic liver cirrhosis, primary biliary cirrhosis, and chronic active hepatitis. Acta Endocrinol Suppl 1983;251:21.
18. Schussler GC, Schaffner F, Korn F: Increased serum thyroid hormone binding and decreased free hormone in chronic active liver disease. N Engl J Med 1980;299:510.
19. Birkhauser M, Busset R, Burer T, et al: Diagnosis of hyperthyroidism when serum thyroxine alone is raised. Lancet 1977;2:53.
20. Gavin LA, Rosenthal M, Cavalieri RR: The diagnostic dilemma of isolated hyperthyroxinemia in acute illness. JAMA 1979;242:251.
21. Kaplan MM, Larsen PR, Crantz FR, et al: Prevalence of abnormal thyroid function test results in patients with acute medical illnesses. Am J Med 1982;72:9.
22. Levy RP, Jensen JB, Laus VG, et al: Serum thyroid hormone abnormalities in psychiatric disease. Metabolism 1981;30:1060.
23. Morley JE, Shafer RB: Thyroid function screening in new psychiatric admission. Arch Intern Med 1982;142:591.
24. Cohen KL, Swigar ME: Thyroid function screening in psychiatric patients. JAMA 1979;242:254.
25. Little JS: Effects of thyroid hormone supplementation on survival after bacterial infection. Endocrinology 1985;117:1431.
26. Melmed S, Geola FL, Reed AW, et al: A comparison of methods for assessing thyroid function in nonthyroidal illness. J Clin Endocrinol Metab 1982;54:300.
27. Chopra IJ, Van Herle AJ, Chua Teco GN, Nguyen AH: Serum free thyroxine in thyroidal and nonthyroidal illnesses: A comparison of measurements by radioimmunoassay, equilibrium dialysis, and free thyroxine index. J Clin Endocrinol Metab 1980;51:135.
28. Slag MF, Morley JE, Elson MK: Free thyroxine levels in critically ill patients: A comparison of currently available assays. JAMA 1981;246:2702.
29. Kaptein EM, MacIntyre SS, Weiner JM, et al: Free thyroxine estimates in nonthyroidal illness: Comparison of eight methods. J Clin Endocrinol Metab 1981;52:1073.
30. Witherspoon LR, Shuler SE, Garcia MM, Zolinger LA: An assessment of methods for the estimation of free thyroxine. J Nucl Med 1980;21:529.
31. Wood DG, Cyrus J, Samols E: Low T_4 and low FT_4I in seriously ill patients: Concise communication. J Nucl Med 1980;21:432.
32. Gow SM, Elder A, Caldwell G, et al: An improved approach to thyroid function testing in patients with nonthyroidal illness. Clin Chem Acta 1986;158:49.
33. Boles JM, Morin JF, Garre MA: Ultrasensitive assay of thyroid stimulating hormone in patients with acute nonthyroidal illness. Clin Endocrinol 1987;27:395.
34. Hasen J, Boczko M, Donovan R, Giannini M: Blunted TSH response to TRH in patients with central nervous system disease. Am J Med Sci 1981;281:130.
35. Rudman D, Fleischer AS, Kutner MH, Raggio JF: Suprahypophyseal hypogonadism and hypothyroidism during prolonged coma after head trauma. J Clin Endocrinol Metab 1977;45:747.

36. Borst GC, Osburne RC, O'Brian JT, et al: Fasting decreases thyrotropin responsiveness to thyrotropin-releasing hormone: A potential cause of misinterpretation of thyroid function tests in the critically ill. J Clin Endocrinol Metab 1983;57:380.
37. Vinik AL, Kalik WJ, McLaren H, et al: Fasting blunts the TRH response to synthetic thyrotropin-releasing hormone (TRH). J Clin Endocrinol Metab 1975;40:509.
38. Wong ET, Bradley SG, Schultz AL: Elevations of thyroid-stimulating hormone during acute nonthyroidal illness. Arch Intern Med 1981;141:873.
39. Sturgess ML, Weeks I, Evans PJ, et al: An immunochemiluminometric assay for serum free thyroxine. Clin Endocrinol 1987;27:383.
40. Spencer CA, Lai-Rosefeld AO, Guttler RB, et al: Thyrotropin secretion in thyrotoxic and thyroxine treated patients: Assessment by a sensitive immunoenzymometric assay. J Clin Endocrinol Metab 1986;63:355.
41. Spratt DI, Pont A, Miller MB: Hyperthyroxinemia in patients with acute psychiatric disorders. Am J Med 1982;73:41.
42. Kramlinger KG, Gharib H, Swanson DW, Maruta T: Normal serum thyroxine values in patients with acute psychiatric illness. Am J Med 1984;76:799.
43. Wood L: Endocrine Soc Program 1980, Abstract 505, p 201.

3

Supersensitive Thyrotropin Assays

IAN D. HAY

Although the in vivo action of pituitary extracts on guinea pig thyroid was first demonstrated in 1929,[1] it was not until the 1950s that bioassays for the measurement of thyrotropin (thyroid-stimulating hormone, TSH) in human plasma were developed. Because of the complex methodology and the limited sensitivity, specificity, and reproducibility, such assays were not generally used by clinicians.[2] In 1965 Odell and co-workers described a radioimmunoassay (RIA) for TSH in human serum[3] and for the next 15 years TSH-RIA methods, with detection limits in the 1 to 2 μU/ml range, were widely applied to the initial diagnosis and subsequent monitoring of patients with primary hypothyroidism.[2]

In 1980 Spencer and Nicoloff[4] described a TSH-RIA that had a functional sensitivity of 0.8 μU/ml, but required a total incubation period of 8 days. Such an assay was one of the first capable of measuring the subnormal TSH levels occurring in hyperthyroid patients, but was too tedious and time consuming for use in routine patient care.[5]

In the past 5 years, several more rapid turnaround assays have become available,[6] based on an alternative approach to immunoassay called immunometric assay (IMA). These newer supersensitive TSH (STSH) assays are capable of distinguishing normal from subnormal serum TSH levels and, thus, have the potential to expand significantly the utility of TSH measurement in clinical practice.[5] Instead of being used only for confirming the diagnosis of primary hypothyroidism and distinguishing primary from secondary (central) hypothyroidism, STSH measurements may soon be considered the single best indicator of abnormal thyroid function[6] and potentially could become the best initial test of thyroid function in all situations of suspected thyroid dysfunction.[5-7]

In this chapter the following aspects of STSH assays are considered: 1) available IMA methods, 2) criteria for analytic performance, 3) criteria for clinical utility, and 4) role in testing for thyroid dysfunction in both ambulatory and hospitalized patients.

Available IMA Methods

In contrast to conventional RIA methods, the newer IMA techniques involve labeling of the antibody component of the antibody-antigen reaction and generally involve producing a "sandwich" or two-site approach, using two separate monoclonal antibodies of differing specificity to react with the antigen molecule. The new STSH assays employ an excess of solid-phase TSH antibody for immunoextraction of serum TSH; after removal of unbound serum constituents, a second labeled TSH antibody is used to allow quantification in terms of the amount of label binding the solid support via the TSH molecules forming a bridge between the two antibodies.[8] To date, the majority of IMAs have used an isotopic label, so-called immunoradiometric assays (IRMA), but, recently, other assays have employed nonisotopic labels that are fluorophor (IFMA), enzymatic (IEMA), or chemiluminescent (ICMA) in nature.

The excess antibody requirement of IMA until recently prevented widespread use of IMAs. However, with the development of monoclonal technology, this limitation has been overcome. Present STSH assays may employ configurations of either monoclonal, polyclonal, or mixed monoclonal-polyclonal TSH or TSHβ antibodies. The advantages of IMA technology and conventional RIA methods have recently been reviewed.[9] They include shortened turnaround time (hours versus days; typically 2- to 4-hour incubation periods), technical simplicity with fewer pipetting steps, increased range (often allowing quantification without dilution up to 200 to 324 μU/ml), improved specificity due to the interaction of two different monoclonal antibodies with the antigen, and improved sensitivity related to the signal-to-noise ratio of the label.[8]

Criteria for Analytic Performance

Since 1983, when the pioneering Hybritech STSH IRMA was described,[10] many new TSH assays have been introduced and most claim to be capable of distinguishing normal from subnormal serum TSH levels. Before, however, all such assays are advocated for general use, they should meet specific strict criteria, both analytical and clinical.[5] In 1987 the Committee on Nomenclature of the American Thyroid Association[6] recommended that "a minimal requirement for an assay of this type is that serum from clinically hyperthyroid patients gives results that are more than 3 log SD below the mean value found in serum from normal subjects. The threshold of measurement of serum TSH (i.e., the 'sensitivity'), and the range of normal serum values should be stated."

At the present time, many manufacturers claim for their assays a "sensitivity" limit that is based on the intraassay precision of duplicates of the zero standard (i.e., the "least detectable dose" approach favored by

Rodbard[11]). Spencer has stated that "this must be considered as a theoretical and often irrelevant number that does not relate to clinical measurements using serum samples."[8] She favors a more realistic "functional sensitivity" limit, defined as the TSH level where the interassay precision is 10% to 15%, a view also favored by Bayer.[12] Such views have, however, not been shared by Hershman et al, who argue that the SD/mean ratio tends to infinity as the dose approaches zero and, thus, the interassay coefficient of variation (CV) "does not provide a uniform measure of assay discrimination between the nondetectable, hyperthyroid and euthyroid ranges."[13]

In 1987 Klee and Hay[5] proposed five performance criteria that they considered generally applicable to STSH assays. It was hoped that use of these criteria could help evaluate the claims made in terms of sensitivity of the newer TSH assay methods. Criterion 1, the sensitivity criterion, required a less than 1% overlap between the variation of the lower normal value limit and the assay detection limit. For the evaluation of the low-end analytical sensitivity, repeated measurements were made on six serum pools (target limits 0 to 0.5 μU/ml) prepared by adding the WHO 68/38 TSH reference standard to zero reference serum purchased from the manufacturers. The assay variances were estimated directly from the pool measurements, rather than using a theoretical model. The assay detection limit was calculated as the TSH value corresponding to the mean plus 2.6 SD of the zero pool radioactivity; the SD of TSH measurement corresponding to the lower normal value used a least squares linear function of the sample SDs of the pools measured. The sensitivity criterion was met if the lower normal value minus 2.6 SD exceeded the assay detection limit.

In a survey of 20 studies published since 1984 on STSH assays using commercially available reagents, where detection limits were stated to be 0.2 to 0.6 μU/ml, Klee and Hay[5] were able to find estimates of the low-end assay variation in only five of the 20 reports, and in six there was no description of the lower normal value limit. Such information should in the future be routinely provided by manufacturers and investigators when describing new TSH assay methods with claims of "improved sensitivity."

Criteria for Clinical Utility

STSH assays capable of detecting subnormal serum TSH concentrations should permit the prediction of response to thyrotropin-releasing hormone (TRH) and the detection of hyperthyroidism.[5] Of Klee and Hay's five performance criteria, the second and third related to the sensitivity and specificity of the tests for predicting patient response to TRH testing. The fourth and fifth criteria related to the ability of STSH assays to classify hyperthyroid patients. The specific performance limits used in these criteria[5] were arbitrarily set, but appeared to function well and effectively discriminated in an evaluation of the two TSH-IRMA assays employed at Mayo, the Hybritech two-step Tandem-R, the Boots-Celltech Sucrosep method, and the more recently introduced Ciba Corning Magic Lite two-site TSH-ICMA method.

To determine the sensitivity and specificity for predicting TRH responsiveness, Klee and Hay's criteria 2 and 3 required subnormal basal TSH values in 95% of patients with subnormal TRH responses, and detectable basal TSH values in 95% of patients with normal TRH responses.[5] In considering TRH stimulation testing in 150 patients with basal STSH values below 1.0 μU/ml, a normal response was considered to be a change in TSH of >2 μU/ml after stimulation. The choice for the two TRH criteria of two different limits (the lower normal for sensitivity, and the detection for specificity) was a deliberate one and allowed for a "gray zone" of unpredictable response in the range of basal STSH values between undetectable and normal (0.06 to 0.39 μU/ml with our assays).

Criteria 4 and 5 required undetectable TSH values in 95% of hyperthyroid patients and detectable values in 95% of clinically euthyroid subjects, "including patients with non-thyroidal disease, but excluding patients taking thyroid replacement therapy."[5] In all but three of 20 published studies evaluated by Klee and Hay, the detection limit was used as a decision value for separating hyperthyroid from euthyroid basal TSH concentrations, and the TSH levels in most clinically hyperthyroid patients were undetectable. The choice of 95% limits allowed for 5% of atypical values, but in practice the criterion of 95% specificity for predicting hyperthyroidism was readily met by all 17 studies that evaluated this criterion, using the assay detection limit as the decision level.[5]

In a recent comparative study of five American STSH assays, Hershman et al[13] found that all hyperthyroid samples were detectable with the DPC Coat-A-Count IRMA and 31% were detectable by the Abbott HTSH IEMA. It was their opinion that criterion 4 may not continue to be appropriate "since advances in TSH assay technology will eventually render all hyperthyroid TSH values detectable."[13] For the present STSH assay market the criterion does, however, appear to be still applicable.

Role in Testing for Thyroid Dysfunction

At present, serum total T_4 and indirect estimates of free T_4 concentration, such as the popular free T_4 index, still represent the most widely used screening tests for thyroid dysfunction.[2,6] It has been argued that the use of STSH assays would be more appropriate since this determination would be less affected by nonthyroidal illness (NTI) and not influenced by changes in thyroid hormone-binding proteins.[7] Such a TSH-based testing strategy would have the advantage of detecting patients with subclinical hyper- or hypothyroidism, in whom by definition total or free thyroid hormones would be normal.[14] However, the effectiveness of such a testing strategy would largely depend on the nature of the clinical setting, such that a TSH screen might only be superior to a T_4 screen in referral centers with a high prevalence of thyroid disorders.[15] Although it has been reported[16] that in hospitalized patients a raised or undetectable TSH result is as likely to be attributable to NTI as to thyroid dysfunction, Toft[14] has emphasized that, at least in the United Kingdom, it is not the practice to screen hospitalized patients indiscriminately for thyroid disease and that, if STSH estimations are restricted to those patients with clinical features, the assay specificity will be significantly improved.

In an early report on the utility of STSH, it was stated that all of 35 patients with NTI had detectable and normal STSH levels.[17] In a later report of 264 consecutive admissions to a general medical unit from the same hospital, STSH levels were undetectable in 1% and raised in 4%. In none of these patients was there evidence of thyroid disease.[18] When 139 patients hospitalized at Mayo Medical Center were screened, 79% had normal STSH, with 10% having low values and 11% having elevated values. However, when the 17 patients with known thyroid disease were excluded, only 2% of patients had subnormal STSH values (undetectable in only one patient), while 9% continued to have unexplained STSH elevations. Although such patients could have had unrecognized hypothyroidism, they were more likely to have been in the recovery phase of severe NTI,[19] when transient rises of TSH to greater than 20 μU/ml have been recorded.[20] Toft has cautioned that "the chance finding of an abnormal TSH result in a patient with NTI should not usually prompt any further action other than remembering to repeat the test after recovery"![14]

In the stable ambulatory patient with intact hypothalamic-pituitary function, who is not receiving drugs known to suppress pituitary TSH secretion, STSH measurement has good clinical sensitivity and specificity for diagnosing thyroid disease.[5,7,21] In such a patient the finding of an abnormal STSH is strongly suggestive of clinical or subclinical thyroid hormone excess or deficiency, which can be confirmed by an indirect estimate of FT_4, or in the case of suspected hyperthyroidism, FT_3.[5,7,14,21] On the other hand, a normal STSH virtually excludes the possibility of thyroid hormone excess or deficiency.

In a hospital setting there appears to be a higher prevalence of abnormal STSH values that are not due to thyroid disease but associated with diverse NTI states, as well as glucocorticoid or dopamine therapy.[16,21] These STSH abnormalities (whether elevated or undetectable) appear to be transient in NTI patients. Thus, in NTI patients with an undetectable STSH level, either repeat STSH measurement or a TRH stimulation test can be valuable to exclude hyperthyroidism.[14,21] NTI patients with an elevated STSH may often be distinguished from sick hypothyroid patients on the basis of a normal T_4 level[16] However, if primary hypothyroidism is suspected, such a diagnosis can usually be confirmed by the findings of a low T_4 and high STSH in a follow-up study after recovery from the NTI.[21]

A TSH-based testing strategy for the evaluation of thyroid function could have many advantages. In the majority of patients, thyroid dysfunction could be satisfactorily excluded with one test. In patients suspected of being hyperthyroid[5,17] or taking thyroid-suppressive treatment for thyroid carcinoma,[22,23] the extra costs, hazards, and inconvenience of TRH stimulation tests could be avoided. Currently, the cost of a STSH assay is considerably more than the cost of either a T_4 or free thyroxine index test.[15] However, if the STSH procedure was automated and the number of specimens tested markedly increased, the cost of a STSH assay could be more competitive. Also, if the traditional lab tests currently used in the follow-up of falsely abnormal primary tests can be largely eliminated, a TSH-based testing strategy for thyroid function assessment should prove to be cost effective.[23]

Discussion by Dr. Joel I. Hamburger

Dr. Hay, have you been impressed by variations in STSH assay values from day to day in patients treated with LT_4, so that one day the STSH value may be normal, and on another occasion (on the same LT_4 dose) the value may be subnormal?

Euthyroid patients sometimes have subnormal STSH values. These patients still need further testing to rule out hyperthyroidism, don't they; or can one conclude that all hyperthyroid patients will have undetectable STSH levels?

Do you still use TRH tests in patients (especially those with goiters of Hashimoto's thyroiditis) who have high normal basal TSH values, and low normal T_4 or FT_4 values, to identify TRH hyperresponders?

Response by Dr. Ian D. Hay

In the 25 months that I have used, on a daily basis, the Boots-Celltech STSH assay, I have not been impressed by variations such as you describe in patients on T_4 therapy. A possible explanation might be that our patients routinely have their bloods drawn in the fasting state between 7 and 9 a.m. Alternatively, the differences you see may be related to the STSH method employed.

Our performance criteria required that undetectable TSH values be found in 95% of hyperthyroid patients and detectable values in 95% of clinically euthyroid subjects. Patients with subnormal, but detectable, STSH values in the 0.1 to 0.39 μU/ml range may often be clinically euthyroid, but in our experience only about one third of these patients will show normal TRH responsiveness and are likely, therefore, to be truly euthyroid. The finding of a subnormal TSH value should always, I believe, "trigger" further testing to rule out thyroid dysfunction.

In 15 years of practice I have not regularly performed TRH tests on patients with low normal T_4 and mildly elevated TSH values. Personally, I would not be more likely to treat such a patient with T_4 if a "hyper-response" to TRH was found.

Discussion by Dr. James C. Sisson

Obtaining a sensitive TSH concentration appears to be a valuable and cost-effective approach to diagnosis in patients suspected of having thyroid dysfunction. But what roles should a priori probabilities of hyperthyroidism and hypothyroidism play in determining which tests to order? For example, any reasonable probability of primary hypothyroidism (e.g., clinical features plus a firm goiter or prior ^{131}I therapy) should be amply confirmed by the STSH (or any TSH) assay. The result will also be a guide to success with therapy.

However, for patients suspected of hyperthyroidism, reaching a decision is more complicated. If the probability of hyperthyroidism is high (possibly over 80%), rather than obtaining a STSH, why not move directly to FT_4 or FT_3 measurements that are likely: 1) to raise the probability of diagnosis to the point where therapy can be given, and 2) to serve as an index of therapeutic response? For lesser probabilities of hyperthyroidism we could choose the STSH at the outset, or a battery of assays if one wishes to obviate return visits for additional tests that may be necessary for confirmation of a low STSH result. Should we not strongly encourage the acquisition of clinical data to establish a priori probabilities as essential to the competent practice of medicine?

When faced with a patient with nonthyroidal illness, the most important question related to the thyroid gland is whether thyroid dysfunction is contributing substantially to the overall sickness of the patient. Particularly vexing is the patient who is desperately ill, and any appropriate action must be soon applied. In such a patient, cannot the TSH concentration at least identify patients who probably manifest advanced primary hypothyroidism so severe as to warrant large doses of thyroid hormone?

Response by Dr. Ian D. Hay

I absolutely agree with Dr. Sisson that the choice of initial thyroid function tests should be dependent on the probability of hyper- or hypothyroidism in the individual patient being evaluated. In patients who are clinically hyperthyroid, I suspect that most U.S. endocrinologists in 1988 are still initially ordering a free thyroid hormone estimate, rather than

STSH measurement. In this context, the current survey being organized by the American Thyroid Association in regard to the diagnosis and treatment of hyperthyroidism in the United States will provide us with much relevant data on current U.S. testing strategies. Certainly, to avoid return visits, it is sometimes necessary to obtain an initial "battery of assays." Hopefully, in the future, the decision to perform appropriate "second-line" in vitro tests might be made at the laboratory level and, thus, further venipuncture and unnecessary delay in making a diagnosis may thereby be avoided.

In a recent study from Stanford of patients with severe NTI,[24] it was confirmed that STSH measurements had "a high diagnostic sensitivity, a relatively poor specificity, and a very low positive predictive value" (as low as 35%) for both hypo- and hyperthyroidism. However, these authors considered that the specificity and predictive value would be vastly improved if TSH values of 0.05 and 12 μU/ml were used as cutoff points, rather than the 0.4 to 6.2 range used for ambulatory patients. Also, they emphasized that *inverse* abnormalities in both STSH and FT_4 were indicative of thyroid dysfunction and that their interpretation in hospitalized patients was "virtually the same as in outpatients." Our experience and that of Spencer et al[16] is consistent with the expressed views of the Stanford group.

Discussion by Dr. Leonard Wartofsky

In your experience, how often have you seen basal TSH levels <1 μU/ml with a maximum TSH response to TRH of <2 μU/ml? What was the diagnostic significance of the results of such cases? Are patients with low basal TSH (0.1 to 0.8) more likely to have a brisk response or a blunted response to TRH?

Response by Dr. Ian D. Hay

In our published study[5] we used the Boots-Celltech STSH IRMA to evaluate the TSH response to TRH in 150 patients with basal TSH less than 1.0 μU/ml. Of these 150 patients, 40 (27%) had a response of >2 μU/ml, 10 (7%) showed a borderline response of 1 to 2 μU/ml, and 100 (67%) had responses below 1.0 μU/ml. No patient in the group of 34 with normal TSH levels of 0.40 to 1.0 μU/ml showed a suppressed response and none of the 92 patients with a basal TSH <0.1 μU/ml showed a normal response. In the group of 24 patients with subnormal (but detectable) levels in the 0.1 to 0.39 μU/ml range, two thirds had responses of <2 μU/ml and one third had normal responses. From this we concluded that: 1) if TSH response was impaired (<2 μU/ml response), the basal TSH would be subnormal (<0.4 μU/ml) in 98% of the cases; and 2) if TSH response was normal, all such patients would have detectable basal values.

From a diagnostic standpoint, we would therefore predict that almost all patients with a basal TSH level >0.4 μU/ml will have a normal TRH response, that patients with basal TSH levels <0.1 μU/ml will likely be suppressed, and that in patients with TSH levels in the 0.1 to 0.4 μU/ml range a normal TRH test may be obtained in only about a third of cases.

On a practical note, we would agree with Lamberg et al[22] that a basal TSH of <0.1 mU/ml "is a sufficient indication of TSH suppression" and thus, in patients with differentiated thyroid cancer on T_4 suppressive therapy, we do not consider that further testing with TRH is regularly indicated. In our opinion, the role of the TRH test (*outside* the evaluation of anterior pituitary function) is now probably limited to further definition of the status of patients whose basal TSH levels are in the subnormal (but detectable) range of 0.1 to 0.39 μU/ml.

Discussion by Dr. John T. Dunn

On page 42, how was hyperthyroidism defined in the studies of undetectable TSH in hyperthyroidism? Can you really be sure that they were all hyperthyroid?

I believe that routine screening for hypothyroidism in adults, particularly older women, should be seriously considered. Do you agree, and do you think it would be cost effective to screen all older women by STSH?

Response by Dr. Ian D. Hay

In our own study we examined the sensitivity of basal STSH for diagnosing hyperthyroidism in 87 patients with untreated Graves disease, whose diagnoses were established on the basis of clinical examination, serum T_4 and T_3 values, 24h RAIU, and the presence of TRAb (measured in an FRTL-5 cell assay). Of the 20 published studies that we reviewed, we could not "really be sure" that all the described patients were truly hyperthyroid.

I agree that serious consideration should be given to routine screening for hypothyroidism in adults, particularly older women. For some years at the Mayo Clinic a serum T_4 has virtually become a routine test for general internists performing periodic medical exams on elderly patients. In the future it is likely that TSH may become a more cost-effective screening tool. However, it should be noted that TSH values progressively increase with age for both men and women. Thus, an upper limit of normal of 6.0 μU/ml would be appropriate for age 40 years, but thereafter the upper limit increases by approximately 1.0 μU per milliliter-decade to nearer 10 μU/ml at age 80 years. Accordingly, appropriate cutoff values would have to be selected if TSH is to be used as a "front line" test in an elderly population.

Discussion by Dr. John E. Freitas

We have found the STSH values to be extremely helpful in quickly resolving questions of possible hyperthyroidism in preoperative patients with hyperthyroxinemia. Readily detectable STSH values obviate unnecessary postponements of scheduled procedures. With automation and greater specimen volume, our assay is becoming very cost competitive.

The upper limit of the normal range for our STSH assay was determined from the mean plus 2 SD of a euthyroid volunteer pool. Is this the best way to define the upper normal limit of the STSH assay?

Response by Dr. Ian D. Hay

Like Dr. Freitas, we have found that a "readily detectable" (euthyroid) STSH value (like a normal TRH response, in former days) aids in the exclusion of thyrotoxicosis in patients with unexplained hyperthyroxinemia, especially in the hospital setting.

The normal values for assays generally are set as the 2.5 and 97.5 percentile limits adjusted for age and sex differences. In assessing our normal range for STSH we used 149 normal subjects judged to be clinically healthy and free of diseases known to affect thyroid metabolism. We found that the distribution of our STSH measurements was non-Gaussian and, thus, the data were transformed logarithmically to bring the residual distributions closer to Gaussian. We also found that the TSH levels progressively increased with age, although there was no apparent male-female difference. In displaying our normal range we therefore considered the upper normal limit to be best defined by the curve corresponding to the geometric age regression plus 2 SD.

Discussion by Mila Bednarz

In Dr. Hamburger's laboratory, we find that supersensitive TSH assays reliably differentiate between normal and subnormal levels of TSH; but the differentiation between undetectable (and thus TRH nonresponsive levels that are compatible with hyperthyroidism) and subnormal (but TRH responsive and thus not compatible with hyperthyroidism) is imprecise.

Dr. Wartofsky has alluded to immunochemiluminometric assays and I would like to know if any of the panel members use this method of assay for TSH, and whether it permits the reliable separation of subnormal from undetectable TSH levels.

Discussion by Dr. James C. Sisson

Dr. James Smart, head of the University of Michigan Ligand Laboratory, has done comparison studies of the immunochemiluminometric assay with the supersensitive TSH assays, and he has kindly agreed to provide us with his data on the subject.

Commentary by Dr. James B. Smart and Dr. Barry G. England: A Chemiluminescence "Sensitive" TSH Assay: How Does It Measure Up?

Almost since the development of immunoassays there have been continuous efforts to replace the radiolabel with nonisotopic labels. This effort has been successful with specific applications but radiolabels remain the primary tag. A new candidate has presented itself in the Ciba-Corning Magic Lite TSH assay. In addition to the use of a chemiluminescent label, this assay is also promoted as a "sensitive" TSH assay, making it a potential candidate for a one-test thyroid screen. This assay is really only new to the U.S. market as it has been marketed in Europe for about 3 years. Thus, this presents the opportunity to evaluate both the chemiluminescent label and a "sensitive" TSH assay.

Principal and Procedure Chemiluminescence[25,26] has been a topic of the literature for over 10 years but this appears to be the first successful commercialization. The Magic Lite TSH assay, which employs a chemiluminescent acridinium ester,[25] is very similar to the older Magic TSH assay in that it is a sandwich assay utilizing paramagnetic particles, to which are bound the capture antibody. The difference is that the labeled antibody is labeled with an acridinium ester, which in the presence of peroxide at high pH emits light photons. Thus the assay is read, after the final separation (accomplished magnetically), by placing the sample into a luminometer, which injects a peroxide and base solution followed by recording the intensity of the light signal produced. Since the actual signal is from emitted photons, the readout is in counts using a photomultiplier (PM) tube. Once the sample has been read it cannot be reread as the acridinium ester is hydrolyzed off of the antibody (Fig. 3.1).

The protocol for this assay is identical to the Ciba-Corning Magic TSH assay, which is an IRMA assay. One hundred microliters of sample and 100 μl of "lite" reagent are incubated 2 hours at room temperature (our evaluation also included the option of an overnight incubation for this step, vide infra). After the initial incubation, 500 μl of Magic particles are added and a second incubation of 30 minutes at room temperature follows. Finally, using a magnetic separator, the solid particles are separated from the liquid phase, 100 μl of water are added, and the sample is read in the luminometer. The luminometer that we used is similar to a gamma-counter in that it contained a movable sample holder chain that would hold up to 300 samples. Thus from the technicians view the only difference is in the final reading, where the samples would normally be placed in a gamma-counter, they are instead placed in a luminometer with the response in both instances read out as counts.

Assay Sensitivity This brings us to the question of actual magnitude of the response observed with a chemiluminescent label. This is of particular

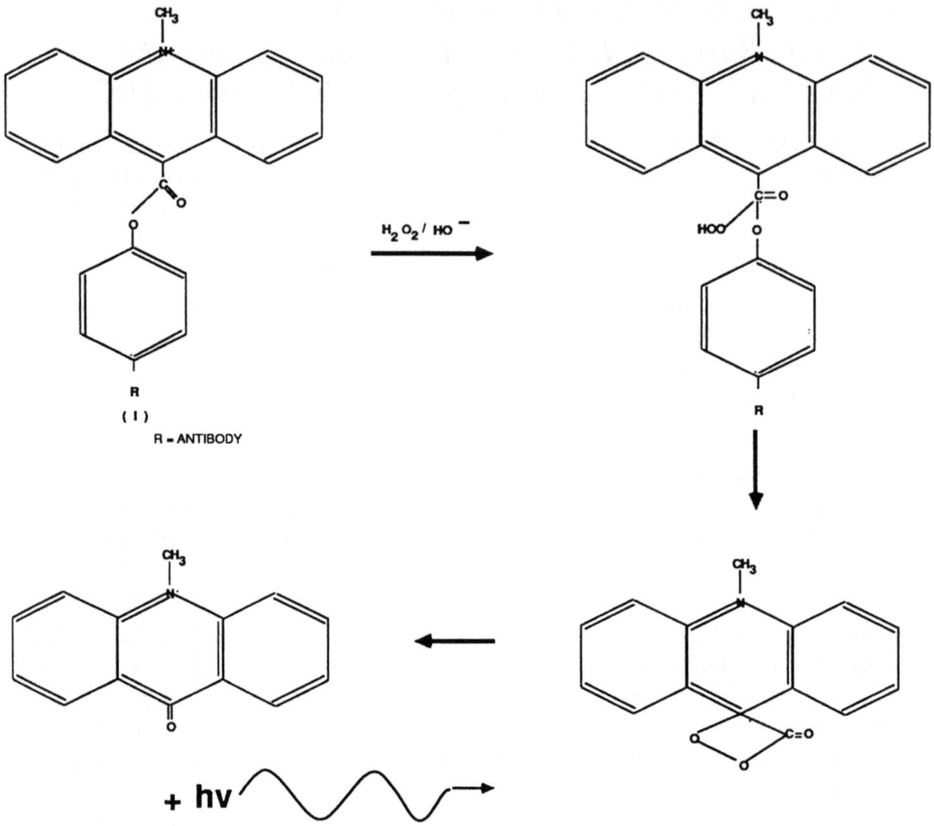

FIGURE 3.1. Chemiluminescent reaction of an acridinium-labeled antibody.

interest in the instance of an IRMA assay where the dividing line between normal and abnormal is often only a few hundred counts per minute. And, even with the generally good reproducibility that the commercial IRMA assays deliver, the low count rates observed in the low end of these assays is a bit discomforting for the laboratory worker. The chemiluminescent label really "shines" in this area with the actual counts observed being at least two orders of magnitude greater than observed with the Nichols radiolabeled assay (Fig. 3.2, a & b).

Why such a dramatic difference? With a radioisotope tag we are looking at only a very small percentage of the labeled atoms since only 0.0006% of them actually decay and emit a detectable gamma-ray in the usual 1 minute counting time allotted a sample. In contrast, 100% of the chemiluminescent-labeled molecules all are activated and emit a detectable photon in about 1 second. This has several advantages: 1) a much larger response signal; 2) the possibility of using either labels of a lower "specific

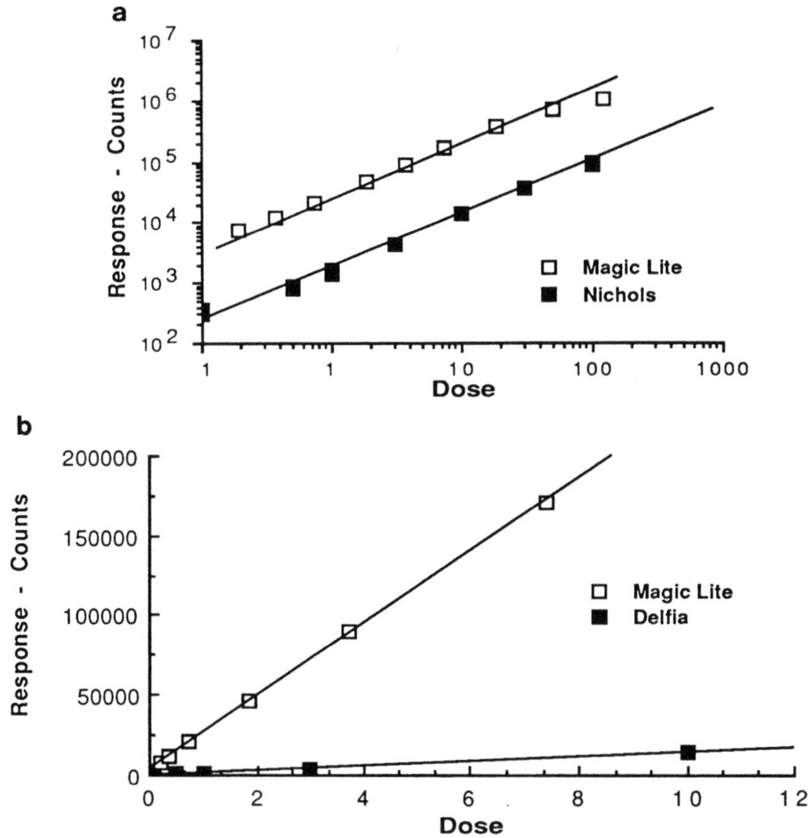

FIGURE 3.2. Dose-response curves for the Magic Lite & Allegro TSH assays. a, Full curve; b, low end (0–2).

activity" or significantly less label, which in theory should permit the assay to respond to a much lower level of analyte; and 3) short counting times (of the order of a few seconds). In addition, if one knows the efficiency of their photomultiplier (PM) tube, the determination of the specific activity or percentage of actually labeled molecules becomes a trivial calculation.

Another feature of this assay is shown in Figure 3.2a, which shows the response curve to be linear to about 100 μU/ml, greatly reducing the number of dilutions and repeat samples that must be run. This is not a feature that is unique to this particular chemiluminescence assay, but in our evaluation of several "sensitive" TSH assays seems to be the rule rather than the exception.

Having commented about the high end of the curve, we might now turn our attention to the low end of the curve and ask of it how "sensitive" is the assay. There has been considerable attention in the literature to the definition of "sensitive" so caution is necessary when using this term. In this particular instance, we are concerned with the least detectable dose that can be distinguished from zero analyte. There are mathematical treatments showing that this can be expressed as a function of curve slope and reproducibility at zero analyte concentration. This is equated with a number by running a zero or buffer sample several times to determine the reproducibility of the assay at this level, then from a curve extrapolated to zero determine the dose value of 2 or 3 SD. This then becomes the least detectable dose and is generally what most manufacturers report as the sensitivity of the assay. In reality, this says that the assay will respond in a detectable manner to any level of TSH present and is only limited by the precision of the assay. It is interesting that all this is accomplished without ever introducing any endogenous TSH into the assay! From a pragmatic view, a function approach seems to be more satisfying in determining this critical number. This can readily be approximated through simple dilution studies (Fig. 3.3). This is not perfect either in that change in matrices becomes a factor along with change in TSH concentration.

Using this as a criterion, the assay limit as demonstrated in Figure 3.3 is somewhere between 0.15 and 0.20. The manufacturer suggests 0.3 as the decision level for hyperthyroid and our studies tend to support this value. Our determination of the sensitivity is well within this value.

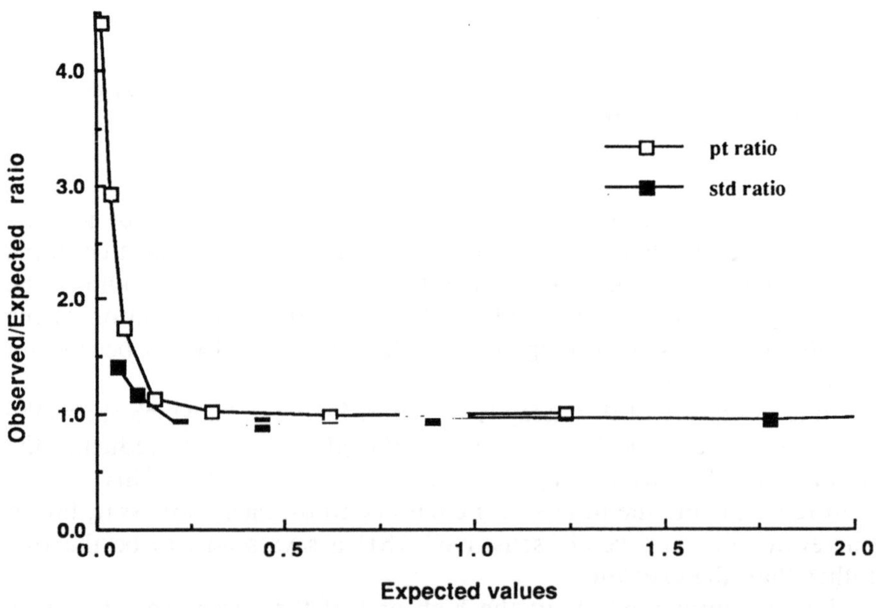

FIGURE 3.3. Dilution studies.

Precision. Figure 3.4 shows the intraassay precision profile of this assay, again compared to the Nichols assay. As with the dose-response comparison, both the profile for the entire working range and the low end only (0 to 10) are shown.

As can be seen, the profiles are comparable and the 15% coefficient of variation (CV) point is below 0.2, which is consistent with the current descriptions of "sensitive" TSH assays.[12] This precision profile is an indication of the intraassay variation. A general estimate of the interassay variation is shown in Table 3.1, which summarizes the data from three controls run over several assays. Unfortunately, data for the critical values around 0.3 μU/ml are not available at this writing, but at the levels compared, the two assays are not remarkably different.

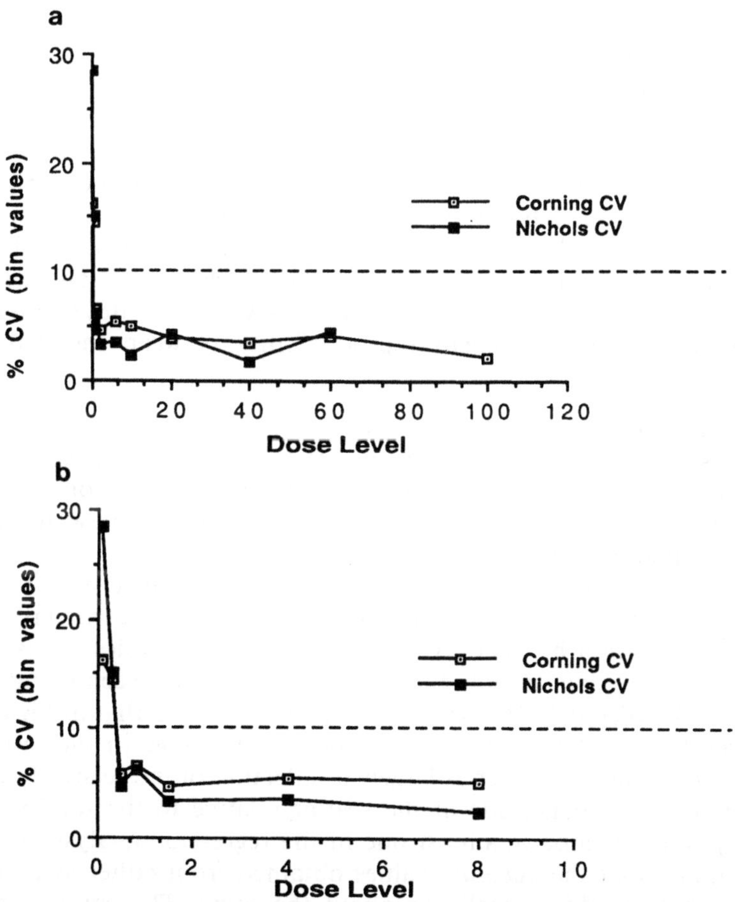

FIGURE 3.4. Precision profile comparison of the Magic Lite and Nichols assays. a, Full range; b, 0–10.

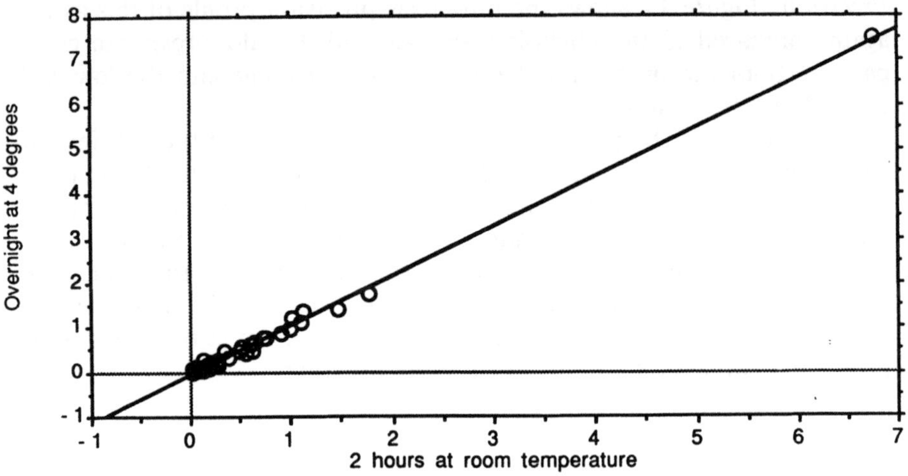

FIGURE 3.5. Correlation between incubations for 2 hours at room temperature and overnight at 4°C. $y = 1.107x - 0.045$; $R^2 = 0.995$; $n = 37$.

For convenience we prefer to run the initial incubation overnight (16 to 20 hours) in our laboratory. Since the manufacturer did not indicate this as an option in their product insert, we compared sera values obtained on several runs under both the protocol described above and with the first incubation carried out overnight at 4 to 8°C. The results, summarized in Figure 3.5 show excellent reproducibility between the two assay protocols.

Accuracy. Having established the precision parameters for the Magic Lite TSH assay, the next order of business is to evaluate the accuracy. One measure is to compare the results from patient sera run on the Magic Lite TSH assay and the Nichols Allegro TSH assay. These results are shown in Figures 3.6 and 3.7.

While Figure 3.6 shows the regression over all patients tested, it is more of academic interest at this point since the primary concern here is with the discrimination between hyper- and euthyroid samples. Thus Figure 3.7 restricts the data to only those samples falling between 0 and 2.0 in the Nichols Allegro TSH assay. The slope is similar to that in Figure 3.6, differing by only 0.096. There is a noticeable change in the respective correlation constant, which is largely a reflection of the relatively greater scatter (i.e., the magnitude of the scatter relative to the total range of values). With a slope of almost one in the regression analysis, it would be expected that the absolute values obtained from either assay on the same serum would generally be about the same. The controls do not follow this dictum, as shown on Table 3.1, and instead the Nichols Allegro TSH assay values are significantly higher than those obtained for the

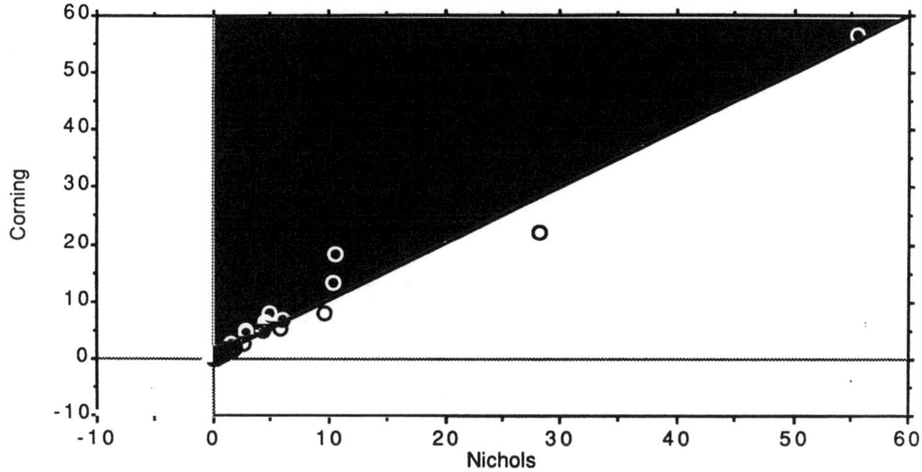

FIGURE 3.6. Correlation between the Magic Lite TSH and Allegro TSH assays. $y = 0.995x + 0.287$; $R^2 = 0.968$; $n = 70$.

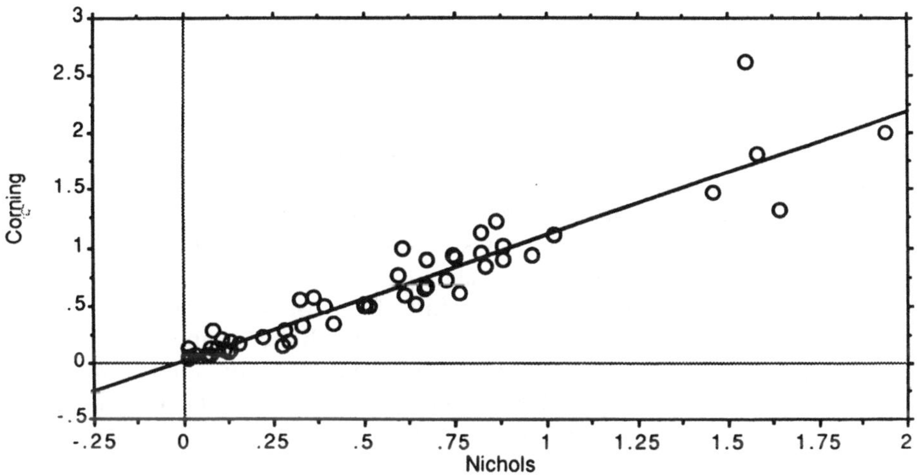

FIGURE 3.7. Correlation between the Magic Lite TSH and Nichols TSH assays in the range 0–2.0 μU/ml. $y = 1.091x + 0.023$; $R^2 = 0.891$; $n = 55$.

corresponding Magic Lite TSH assay. This would suggest a different response to the matrix of the control (these are lyophilized controls) by the two assays. However, both sets of values are within the range of values published for these controls. Inclusion of these controls in our laboratory on assays of other manufacturers generally tends to yield values closer to the lower Magic Lite TSH values.

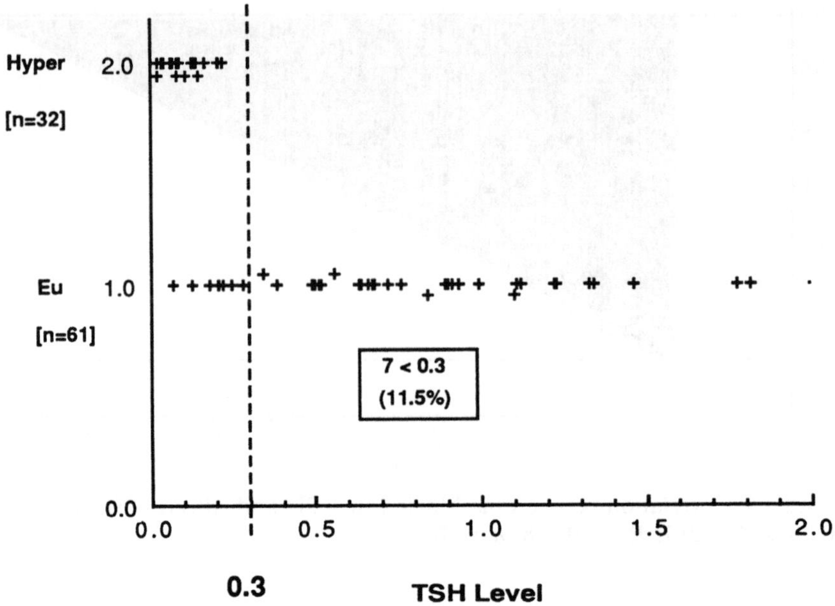

FIGURE 3.8. Histogram for hyper- and euthyroid patients—Allegro TSH assay.

Of course, the real measure of any assay is its diagnostic accuracy. Histograms for both assays in the hyper- and euthyroid ranges have been constructed and are shown in Figures 3.8 and 3.9. From these histograms, the ability of both assays to delineate hyperthyroidism is very good, with no false negatives in the population studied. The number of false positives is significant. Most of these are related to the patients' actual clinical status. Specifically, with the Magic Lite TSH assay, of the seven false positives (i.e. <0.3), two were on Synthroid and diabetic, one was diabetic with other nonthyroid illness (NTI), two were diabetic only, and two were NTI only. In the false positives observed on the Nichols Allegro TSH assay (<0.4 μl/ml was used, but <0.3 μl/ml "looks" better in this study),

TABLE 3.1. Comparison of quality control sera data.

TSH Assay	QC Level	Mean	Intraassay CV	Interassay CV
Corning	Low	1.28	3.14%	2.88%
	Medium	7.60	4.74%	3.90%
	High	22.53	3.08%	4.79%
Nichols	Low	1.57	2.30%	5.72%
	Medium	9.73	0.97%	2.90%
	High	28.45	1.58%	2.79%

QC = quality control.

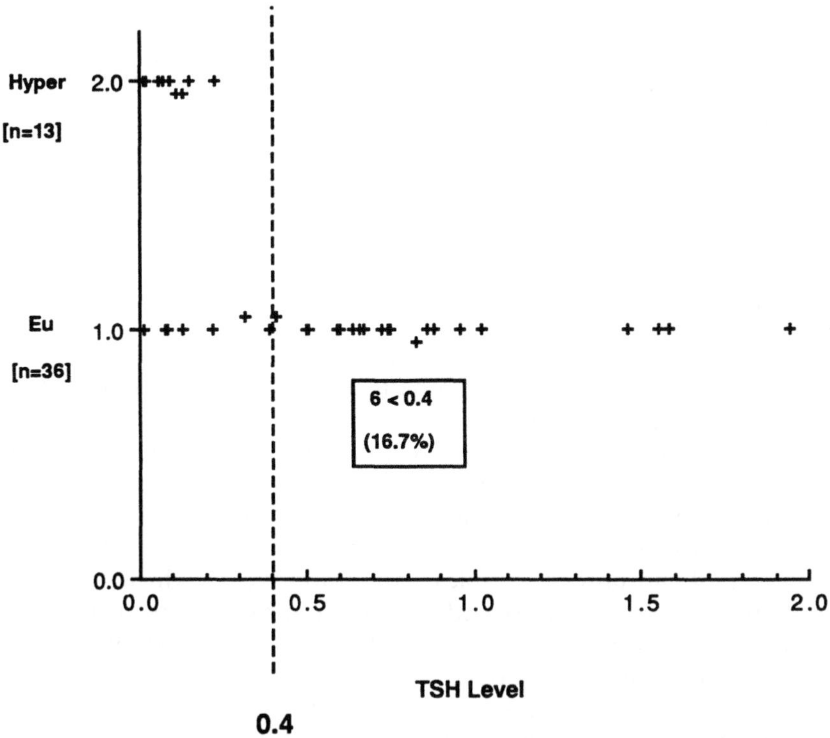

FIGURE 3.9. Histogram for hyper- and euthyroid patients—Magic Lite TSH assay.

two were on Synthroid and diabetic (same two observed in the Magic Lite TSH assay), one has undergone ^{131}I treatment for hyperthyroidism, and one had other NTI.

Therefore, in this study by observing just the TSH values, a significant number of samples (10% or greater) may give false positive results. While further study of the patients in question usually can explain the results, there is sufficient doubt at this time to preclude the use of the "sensitive" TSH assay as a single thyroid screen.

Advantages and Disadvantages. Considering only the basic methodology of the Magic Lite TSH assay, there are several pros and cons that have been observed in our laboratory. First, the advantages:

1. The use of a nonradiometric label greatly reduces the waste disposal problems and most of the problems unique to radiation safety in the laboratory. This also allows for bulk packaging of the reagent.
2. Extended shelf life (6 months or greater).
3. Decreased reading (counting) time: about 6 seconds per sample.
4. Requires only two standard "calibrators."

5. Large signal-to-background ratio, even at low doses.
6. The procedure is essentially identical to the corresponding Magic TSH assay and thus is no more tech intensive than a typical IRMA assay.
7. A proportionately greater measurable response at all dose levels. This permits the use of much lower concentrations of labeled antigen.
8. Low background—there are not many naturally occurring chemiluminescent compounds and deterioration of label usually does not produce chemiluminescent product.

Disadvantages that we have encountered are:

1. The use of a stored curve that is modified by the two calibrators in the assay greatly reduces the amount of quality control data that can be accumulated on the curve.
2. Each tube can be read only once (the label is dissociated from the parent molecule). We experienced some difficulty with the luminometer chain breaking early in the evaluation. This resulted in the loss of data from the beginning of the run, which included our calibrators, leaving the rest of the unread samples with no curve to which to refer to determine the dose. We solved the problem by running a set of calibrators at the beginning and end of the assay. The chain problem was resolved and has not recurred.
3. The label (and acridinium ester derivative) is about the size of two to three iodide ions and thus might interfere with assays for small antigens.
4. New instrumentation is required (a luminometer).
5. Currently there are a small number of assays available employing the chemiluminescent label.

There are two other potential problems of which we are aware that other labs have encountered: scratched tubes (plastic tubes are recommended) that give a high background and latex gloves, which when rubbed on the surface of a plastic test tube give them a static charge that interferes with the reading of the final mixture. We have not experienced any difficulties with either of these situations in our laboratory.

Summary. We have reviewed our experience with the Ciba-Corning Magic Lite TSH assay from two reference points, the use of a chemiluminescent label in place of a radioisotope and its performance as a "sensitive" TSH assay. In both cases we were favorably impressed. The advantages of using a nonradioisotope label amply outweighed any disadvantages, and it is no more demanding of tech time than a routine radioassay.

As a "sensitive" TSH assay, the performance was certainly equivalent to the reference IRMA assay that we employed. The significant increase in response signal in the low end over the IRMA methodology represents a distinct advantage. It is apparent that the use of chemiluminescent labels presents a viable alternative to the use of radioisotopes in many assays.

Acknowledgments. The authors would like to acknowledge M.J. MCully and D.A. Tue who performed most of the assays from which the data in this commentary were obtained.

Commentary by Mila Bednarz

Our comparison of the performance of the Corning Magic Lite TSH assay (ICMA method) with our standard assay, Hybritech's high-sensitivity immunoradiometric assay, was designed to evaluate the reliability of these assays in the detection of low levels of TSH. Almost any modern monoclonal TSH assay is adequate for the detection of elevated TSH values. The accurate measurement of low concentration of TSH by immunoradiometric assay is limited by the rising standard error of the test resulting from the very low count rates at the lower end of the scale. Since the Magic Lite procedure does not involve assay for radioactivity, the possibility of greater assay precision at the lower end of the scan was suggested.

Our first testing of these methods compared the intraassay variation on our low-value pooled serum control. The mean reading with the Hybritech method ($N = 15$) was 0.194 μU/ml ± 0.148 (2 SD), for a CV of 76.3%. The comparable value for Magic Lite ($N = 16$) was 0.187 μU/ml ± 0.038, for a CV of 20.1%. Interassay precision for Hybritech ($N = 19$) gave a CV of 86.6%, while the corresponding figure for Magic Lite was 25.7%.

The mean of the smallest single value that could be distinguished from zero ($N = 20$) was 0.011 ± 0.0374 μU/ml for Magic Lite, and 0.0284 ± 0.092 μU/ml for Hybritech.

Dilution of our normal serum pool was carried out to 1:128 and gave the findings shown in Table 3.2 (studies done in duplicate).

Clearly, the sensitivity of the Hybritech assay was far less as the concentrations of TSH fell below 0.5 μU/ml. The correlation coefficient

TABLE 3.2 Comparison of Magic Lite and Hybritech TSH assays on dilutions of pooled serum from normal patients.

Dilution	Magic Lite (μU/ml)	Hybritech (μU/ml)
undiluted	2.3	2.0
1:1	1.2	1.08
1:2	0.6	0.63
1:4	0.3	0.66/01
1:8	0.15	ND
1:16	0.06	ND
1:32	0.02	ND
1:64	0.01	ND
1:128	0.005/ND	ND

ND = not detected.

($N = 77$) for the two tests was 0.984; the slope was 1.264 and the intercept was −0.56. Because of the negative intercept, the line is skewed at the low end indicating that values below 0.4 μU/ml do not fall on the line. This provides further evidence of different sensitivities for the two tests at extremely low TSH concentrations.

All 18 hyperthyroid patients had undetectable TSH levels by Magic Lite; but levels of 0. 1μU/ml were obtained on 7 of the 18 patients by the Hybritech method.

These studies clearly show that the Corning Magic Lite assay has better sensitivity and reproducibility when TSH concentrations are low. The hands-on time for the two assays are comparable. Further evaluation is necessary before it can safely be concluded that all patients with hyperthyroidism, as well as those on overdosage of thyroxine, will have undetectable TSH values by Magic Lite. It is also necessary to determine the level of TSH, as detected by Magic Lite, below which a patient will be unresponsive to TRH.

References

1. Loeb L, Bassett RB: Effect of hormones of anterior pituitary on thyroid gland in the guinea pig. Proc Soc Exp Biol Med 1929;26:860.
2. Gruhn JG, Barsano CP, Kumar Y: The development of tests of thyroid function. Arch Pathol Lab Med 1987;111:84.
3. Odell WD, Wilbur JF, Paul WE: Radioimmunoassay of human TSH in serum. Metabolism 1965;25:1179.
4. Spencer CA, Nicoloff JT: Improved radioimmunoassay for human TSH. Clin Chem Acta 1980;108:415.
5. Klee GG, Hay ID: Sensitive thyrotropin assays: analytic and clinical performance criteria. Mayo Clin Proc 1988;63:1123.
6. Larsen PR, Alexander NM, Chopra IJ, et al: Revised nomenclature for tests of thyroid hormones and thyroid-related proteins in serum. Arch Pathol Lab Med 1987;111:1141.
7. Caldwell G, Kellett HA, Gow SM, et al: A new strategy for thyroid function testing. Lancet 1985;1:1117.
8. Spencer CA: Clinical utility of sensitive TSH assays. Thyroid Today 1986;9:1.
9. Al-Shawi A, Ali M, Houts T: Principles of labeled antibody immunoassays. Ligand Q 1981;4:43.
10. Miller PVH, Myrtle JF, Wang R: TSH quantification in human serum by a sensitive and specific immunoradiometric assay. Clin Chem 1983;29:1160.
11. Rodbard D: Statistical estimation of the minimal detectable concentration ("sensitivity") for radioligand assays. Anal Biochem 1978;90:1.
12. Bayer MF: Performance criteria for appropriate characterization of "(highly) sensitive" thyrotropin assays. Clin Chem 1987;33:630.
13. Hershman JM, Pekary AE, Smith VP, Hershman JD: Evaluation of five high sensitivity American TSH assays. Mayo Clinic Proc 1988;68:1133.
14. Toft AD: Use of sensitive immunoradiometric assay for thyrotropin in clinical practice. Mayo Clinic Proc 1988;63:1035.

15. Beck JR: Laboratory decision science applied to chemometrics: strategic testing of thyroid function. Clin Chem 1986;32:1707.
16. Spencer CA, Eigen A, Shen D, et al: Specificity of sensitive assays of thyrotropin (TSH) used to screen for thyroid disease in hospitalized patients. Clin Chem 1987;33:1391.
17. Seth J, Kellett HA, Caldwell G, et al: A sensitive immunoradiometric assay for serum thyroid stimulating hormone: A replacement for the thyrotropin releasing hormone test? Br Med J 1984;289:1334.
18. Gow SM, Elder A, Caldwell G, et al: An improved approach to thyroid function testing in patients with non-thyroidal illness. Clin Chem Acta 1986;158:44.
19. Hamblin S, Dyer SA, Mohr VS, et al: Relationship between thyrotropin and thyroxine changes during recovery from severe hypothyroxinemia of critical illness. J Clin Endocrinol Metab 1986;62:717.
20. Brent GA, Hershman JM, Braunstein GD: Patients with severe nonthyroidal illness and serum thyrotropin concentrations in the hypothyroid range. Am J Med 1986;81:463.
21. Spencer CA: Clinical utility and cost-effectiveness of sensitive thyrotropin assays in ambulatory and hospitalized patients. Mayo Clinic Proc 1988;63:1214.
22. Lamberg BA, Helenius T, Liewendahl K: Assessment of thyroxine suppression in thyroid carcinoma patients with a sensitive immunoradiometric assay. Clin Endocrinol (Oxf) 1986;25:259.
23. Bartalena L, Martino E, Falcone M, et al: Evaluation of the nocturnal serum thyrotropin (TSH) surge, as assessed by TSH ultrasensitive assay, in patients receiving long-term L-thyroxine suppression therapy and in patients with various thyroid disorders. J Clin Endocrinol Metab 1987;65:1265.
24. Bayer MF, Macoviak JA, McDougall IR: Diagnostic performance of sensitive measurements of serum thyrotropin during severe non-thyroidal illness: their role in the diagnosis of hyperthyroidism. Clin Chem 1987;33:2178.
25. Weeks I, Beheshti I, McCapra F, et al: Acridinium esters as high-specific-activity labels in immunoassay. Clin Chem 1983;29:1474.
26. Weeks I, Woodhead JS: Chemiluminescence immunoassay. J Clin Immunoassay 1984;7:82.

4

Strategies for Cost-Effective Thyroid Function Testing with Modern Methods

JOEL I. HAMBURGER

The evolution of in vitro thyroid function tests toward ever more sensitive and specific assays has been truly remarkable. This has involved the translation of sophisticated scientific research into simple kit methods for the measurement of biologically active substances present in blood in minute quantities. However, with such dynamic progress, there is inevitably the potential for confusion. The multiplicity of thyroid function tests available creates problems for the nonthyroidologist physician who usually just wants to know whether his or her patient has hypothyroidism, hyperthyroidism, or neither. Which tests are best? Unfortunately, for various reasons normal patients often have abnormal results with commonly employed thyroid function tests.[1,2] A battery of tests is not only inefficient, but often inconclusive when data are conflicting.

This chapter shows how newer test methods reduce the potential for diagnostic errors, and presents a clinically oriented, cost-effective approach to the laboratory confirmation of clinical diagnoses of thyroid dysfunction.

In the further discussion the terms and definitions defined in Table 4.1 will be used.

Thyroid Physiology

An understanding of the essentials of thyroid physiology is important if one hopes to avoid pitfalls in the use of thyroid function tests. Thyroid function is controlled by the pituitary in the familiar negative feedback regulation format. Thyroid-stimulating hormone (TSH) promotes synthesis and secretion of thyroid hormone as needed. When circulating thyroid hormone levels are insufficient, the serum TSH level rises. When circulating thyroid hormone levels are too high, the serum TSH level is suppressed. The thyroid gland synthesizes and secretes principally thyroxine (T_4), but also lesser amounts of triiodothyronine (T_3).[3] The clearance of T_3 from the blood is much more rapid than that of T_4, so that

TABLE 4.1. Abbreviations and definitions.

FT_3:	Free T_3, or assay for same
FT_4:	Free T_4, or assay for same
FTI:	Free thyroxine index, calculated as the numerical product of values for the T_4 and T_3RU
IRMA:	Immunoradiometric assay
RIA:	Radioimmunoassay
RAIU:	Radioactive iodine uptake, the 24 hour value (although values at any time between 12 and 96 hours, even if slightly higher or lower, are of equal usefulness)
STSH:	Supersensitive IRMA assay for TSH
TBG:	Thyroxine-binding globulin
TBPA:	Thyroxine-binding prealbumin
TRH:	Thyrotropin releasing hormone
TRH test:	Assay for TSH before and 20 minutes after intravenous administration of 100 μg of TRH
T_3:	Triiodothyronine, and the RIA for T_3
T_3 suppression test:	RAIU before and after 4–5 days of T_3 administration, 25 μg three times daily
T_4:	Thyroxine, and the RIA for T_4
T_3RU:	T_3 resin uptake
TSH:	Thyroid stimulating hormone

the T_3 secreted by the thyroid plays only a minor role in thyroid function.[4] More than 99% of the T_4 and T_3 in the blood is bound to carrier protein. The principal carrier protein is thyroxine-binding globulin (TBG). TBG binds T_4 and T_3, although binding of T_4 is more vigorous. Thyroxine-binding prealbumin (TBPA) is a protein that almost exclusively binds T_4.[5] Finally, albumin is a weak binding protein for the thyroid hormones; however, an abnormal albumin avidly binds T_4 in the familial dysalbuminemic hyperthyroxinemia syndrome.[6] Bound thyroid hormone is metabolically inactive. It is the tiny unbound or free hormone fraction that correlates well with thyroid function.[7]

Free thyroid hormone, predominantly FT_4, enters the cells of peripheral tissues and is converted to T_3 under the influence of the enzyme 5′-deiodinase. This conversion takes place principally in the liver and kidneys, but also in other tissues.[8] T_3 is about four times as potent metabolically as T_4. Since intracellular nuclear receptors for thyroid hormone have a 10-fold greater affinity for T_3 than for T_4[3] T_3 appears to be the principal thyroid hormone active at the tissue level. The further metabolism of T_3 to the inactive diiodotyrosine also is mediated by 5′-deiodinase. When drugs, illness, or carbohydrate deprivation prevent conversion of T_4 to T_3, the metabolism of T_4 shifts to the production of reverse T_3, an inactive hormone. Many factors may alter this simple scheme so that euthyroid patients may have abnormal thyroid test data, and vice versa.

Thyroid Function Tests

Many of the currently available thyroid function tests have outlived their usefulness and are now obsolete. Others have continuing but limited usefulness, while still others provide the most reliable information. Table 4.2 classifies commonly used thyroid function tests in terms of their diagnostic utility, also indicating what they measure, and the methods. Normal values for my laboratory are provided for tests with ongoing usefulness.

Obsolete Tests

The T_3 resin uptake (T_3RU) has been employed primarily to correct T_4 assay values for the impact of abnormal protein binding. The numerical product of values for the T_3RU and the T_4 assay provides a free thyroxine

TABLE 4.2 Relative usefulness of thyroid function tests.

Tests	What is measured	Method	Normal range
Obsolete tests			
T_3 RU	Protein binding of T_3	Radioassay	*
FTI	Estimates FT_4 concentration	The numerical product of values for the T_4 and T_3RU tests	
FT_4, 1 step	FT_4	RIA, 1 step, analog	
TSH	TSH	RIA	
T_3	Serum T_3 concentration	RIA	
Tests with limited utility			
T_4	T_4 concentration	RIA	4.5–11.5 μg/dl
TBG	TBG concentration	RIA	14-21μg/dl
TRH test	Pituitary TSH release in response to TRH	IRMA	Increase of 2–18 μU/ml over the baseline value
RAIU	Thyroid clearance of iodide	Scintillation counting	10–30%**
Most reliable tests			
FT_4, 2 step	FT_4 concentration	Gammacoat, 2 step	0.6–2.2 μg/dl
FT_3	FT_3 concentration	RIA, magnetic	2–7 pg/dl
STSH	TSH	IRMA	0.3–4 μU/ml

T_3RU = T_3 resin uptake; FTI = free thyroxine index; FT_3 = free triiodothyronine; FT_4 = free thyroxine; TSH = thyroid-stimulating hormone; RIA = radioimmunoassay; TBG = thyroxine-binding globulin, RAIU = radioactive iodine uptake; TRH = thyrotropin-releasing hormone; IRMA = immunoradiometric assay; STSH = supersensitive IRMA assay for TSH.

*Normal ranges are not provided for obsolete tests.

**These values have very low specificity for thyroid function; see Table 4.4.

index (FTI) of the FT_4 concentration.[9,10] Use of the T_3RU, a test that measures T_3 binding, to correct the T_4 value for abnormal protein binding of T_4 seems intuitively illogical, even though it works most of the time. Since proteins specific for T_4 binding are more common than had been appreciated,[6,11–14] the FTI has proved inconsistently reliable as an estimate of FT_4. Direct assay of the FT_4 concentration is now both practical and reasonably reliable,[15] hence the T_3RU and FTI are obsolete.

Conventional TSH assays have been replaced by the supersensitive TSH assay (STSH). Assays of the serum T_3 concentration measure predominantly metabolically inactive protein-bound T_3. Hence, that assay is unreliable in the diagnosis of thyroid function, and has been superceded by a FT_3 assay.[16]

Tests with Limited Utility

Most of the T_4 measured in the T_4 assay is protein bound, and thus both metabolically inactive and irrelevant to the diagnosis of thyroid function. Nevertheless, the T_4 assay is the most direct measurement of serum protein binding capacity for T_4. Elevations of the T_4 in euthyroid patients suggest increased protein binding, and vice versa.[17–25] If there is no obvious explanation for the increased protein binding (e.g., estrogen administration, pregnancy) or decreased protein binding (e.g., androgen administration, genetic), one might want to consider the other causes of abnormal protein binding outlined in Table 4.3. It is not intended that Table 4.3 should provide more than a general indication of well recognized causes. Up-to-date comprehensive coverage is impossible because new drugs or conditions that may alter protein binding are being reported with great regularity.

Free hormone assays provide reliable data even when protein binding is abnormal, hence the detection of abnormal protein binding is less important in the diagnosis of thyroid function. However, that information may be the first indication of a nonthyroidal illness requiring further evaluation or treatment.

The STSH assay has eliminated the need for thyrotropin-releasing hormone (TRH) testing in most cases of suspected hyperthyroidism. TRH testing (assay for TSH before and 20 minutes after intravenous administration of 100 μg of TRH) may still be useful when a subnormal STSH value is obtained, but the clinical evidence of hyperthyroidism is unconvincing. An increased release of TSH in response to TRH has continuing limited usefulness in the diagnosis of early hypothyroidism. The STSH assay is discussed in greater detail subsequently.

The radioactive iodine uptake (RAIU) is not a useful primary test of thyroid function, because any given value can be seen in any state of thyroid function (Table 4.4). The RAIU is important in the differentiation

TABLE 4.3. Principal causes of abnormal protein binding of thyroid hormone.

A. Increased TBG concentration or affinity for thyroid hormones
  1. Physiologic
    a. Pregnancy
    b. Newborn
  2. Genetic
  3. Drug-induced
  4. Nonthyroidal illness
    a. Liver disease
    b. Acute intermittent porphyria
    c. Hydatidiform mole
    d. Estrogen-producing tumors
    e. Lymphosarcoma
B. Decreased TBG concentration or affinity for thyroid hormones.
  1. Genetic
  2. Drug-induced
  3. Nonthyroidal illness
    a. Liver disease
    b. Protein wasting diseases
  c. Euthyroid sick syndrome*
C. Increased TBPA
  1. Genetic
  2. Islet cell tumors
D. Decreased TBPA in the euthyroid sick syndrome*
E. Increased albumin binding (euthyroid dysalbuminemic hyperthyroxinemia)
F. Increased binding of T_3 or T_4 by antibodies
  1. Autoimmune thyroid disease
  2. Waldenstrom's macroglobulinemia
  3. Hepatocellular carcinoma
  4. Sjogren's syndrome

*In the euthyroid sick syndrome, protein binding abnormalities combine with 5′-deiodinase deficiency to produce test abnormalities.

of conventional forms of hyperthyroidism (ie., Graves' disease, toxic autonomous nodules, and toxic multinodular goiter) from the syndromes of hyperthyroidism with low radioiodine uptake (Table 4.5). The RAIU also may be used to detect nonsuppressible thyroid function in patients who develop clinical or laboratory evidence suggesting hyperthyroidism while on treatment with thyroid hormone. These are usually patients treated for goiter, but some patients after radioactive iodine or surgical therapy for hyperthyroidism later develop recovering thyroid function that is not suppressible by thyroxine. An RAIU of more than 5% to 10% in the face of an elevated FT_4 value and a subnormal STSH value signifies an important degree of nonsuppressible thyroid function. The RAIU value also is used in the calculation of radioactive iodine therapy doses. Finally, the RAIU may be useful occasionally as part of a T_3 suppression test (RAIU before and after 4 to 5 days of T_3 administration, 25 μg three times daily).

TABLE 4.4. Nonspecificity of RAIU values for the diagnosis of thyroid function.

A. Elevated RAIU values may be seen in
  1. Hyperthyroidism
  2. Euthyroidism
    a. Rebound after withdrawal of thyroid hormone or antithyroid drugs
    b. Recovery from subacute thyroiditis
    c. Early phases of Hashimoto's thyroiditis
    d. Compensated dyshormonogenesis
    e. Low iodine diet
  3. Hypothyroidism
    a. Decompensated dyshormonogenesis
    b. Hashimoto's thyroiditis

B. Normal RAIU values may be seen in
  a. Hyperthyroidism
    a. The patient is on antithyroid drugs
    b. The patient has received iodides
    c. Toxic multinodular goiter or toxic autonomously functioning thyroid adenoma
  2. Euthyroidism
  3. Hypothyroidism
    a. Hashimoto's thyroiditis
    b. After treatment of hyperthyroidism(^{131}I or thyroidectomy)
    c. Decompensated dyshormonogenesis
    d. In the recovery phase of subacute thyroiditis

C. Low RAIU values may be seen in
  1. Hypothyroidism
  2. Euthyroidism
    a. The patient has received iodides or thyroid hormone (less likely antithyroid drugs)
    b. Hashimoto's thyroiditis
    c. Subsiding subacute thyroiditis
  3. Hyperthyroidism (see Table 4.5)

Most Reliable Tests

FT_4 Assay

The Gammacoat, two-step assay for FT_4 is a simple and relatively reliable assay that gives appropriate values even for patients with dysalbuminemic hyperthyroxinemia, as shown by Braverman's group,[15] and confirmed in my laboratory. It does have a rather high coefficient of variation, so any given value may differ from the true value by up to 10%. Most commercial FT_4 kit assays are one-step analog procedures. These methods give falsely elevated values in patients with increased T_4 binding by proteins other than TBG (as does the FTI), and are not recommended.[26–28].

FT_3 Assay

A simple and reliable assay for the FT_3 concentration is also available (Amersham, Arlington, IL).[16] To avoid interference by T_3 binding antibodies it is essential to add polyethylene glycol to the assay.[24,25] The

TABLE 4.5. Causes of hyperthyroidism with low RAIU.

A. Common causes
1. Subacute granulomatous thyroiditis (painful thyroiditis)
2. Subacute lymphocytic thyroiditis (painless thyroiditis)
3. Jodbasedow disease (iodide-induced hyperthyroidism)
4. Iatrogenic or factitious thyrotoxicosis

B. Less common causes
1. After radiation therapy to the neck (e.g., for lymphoma)
2. Metastatic malignancy to the thyroid
3. Acute hemorrhagic infarction of an autonomously functioning thyroid adenoma
4. Toxic ovarian struma
5. Massive metastases of thyroid cancer outside the neck
6. After adrenalectomy for Cushing disease

principal use for the FT_3 assay is in the diagnosis of mild hyperthyroidism, especially T_3 toxicosis (i.e., hyperthyroidism without an elevation in the T_4 value). T_3 toxicosis is seen in early stages of Graves disease, toxic autonomous nodules, and less often toxic multinodular goiter. Assays for FT_3 are useless in the diagnosis of hypothyroidism.

Supersensitive TSH Assay

Radioimmunoassay for TSH initially was useful primarily because high values confirmed diagnoses of primary hypothyroidism, or inadequate treatment of that disease. Until recently it was not possible to differentiate subnormal from normal values. Supersensitive immunoradiometric TSH assays (STSH) permit such a discrimination.[29-31] Subnormal STSH levels may be seen with hypopituitarism, although most of these patients have normal STSH values. More common is the need to recognize that circulating thyroid hormone levels are excessive, especially when other clinical and laboratory evidence for hyperthyroidism is inconclusive. Thus a subnormal STSH value suggests hyperthyroidism. Perhaps equally important, a normal STSH value virtually excludes hyperthyroidism.

A subnormal STSH value, although compatible with hyperthyroidism, is not diagnostic of that condition. This finding can be seen in patients with euthyroid stages of Graves disease, autonomous nodules, and multinodular goiters. Perhaps these patients, even though not clinically or biochemically hyperthyroid, still have oversupplies of thyroid hormone for them. Nevertheless, they do not necessarily progress to overt thyrotoxicosis. Subnormal STSH values also may be seen in euthyroid patients who are receiving adrenal steroids, or dopamine, or who are malnourished, depressed, or acutely ill. After TRH administration in these patients, STSH levels usually rise significantly, although subnormally.

Current enthusiasm for STSH assays has led to the idea that the STSH test might be used as the primary test for thyroid function,[29] high values

TABLE 4.6. Causes of abnormal thyroid function test data other than thyroid dysfunction and abnormal protein binding.

A. Deficiency of 5′-deiodinase
   1. Drug related
   2. Familial
B. Combined abnormal protein binding and 5′-deiodinase deficiency in the euthyroid sick syndrome.
C. Miscellaneous nonthyroidal conditions
   1. Resistance to thyroid hormone
      a. Generalized
      b. Limited to the pituitary
      c. Limited to the periphery
   2. Acute psychosis
   3. Hyperemesis gravidarum
D. Drugs
   1. Thyroid hormone
   2. Amphetamines
   3. Heparin

meaning hypothyroidism, low values suggesting hyperthyroidism, and normal values excluding thyroid dysfunction. This may be cost effective when screening for thyroid dysfunction. In the diagnosis and management of thyroid disease, or in the exclusion of thyroid disease in the difficult situations discussed below, assay of the free thyroid hormone levels, and sometimes other tests, will be needed.

Other Causes of Abnormal Thyroid Function Test Data in Euthyroid Patients

The principal causes of abnormal thyroid function test data in euthyroid patients, other than abnormal protein binding are outlined in Table 4.6. These situations produce seemingly conflicting results for thyroid function tests. Common patterns are cited in Table 4.7, along with diagnostic considerations and useful confirmatory tests. The following discussion amplifies the outlines of Tables 4.6 and 4.7.

Deficiency of 5′-Deiodinase

Deficient 5′-deiodinase activity is most common in the euthyroid sick syndromes.[1,32–36] However, these patients also have binding protein abnormalities that compound the problem of function test interpretation. Hence, the euthyroid sick syndromes will be dealt with separately. Isolated 5′-deiodinase deficiencies are usually drug related, but may be familial.[37] The defect reduces the efficiency with which T_4 is converted to T_3. T_3 and FT_3 levels are depressed, but T_4 and FT_4 levels are increased because of the decreased deiodination of T_4. STSH levels are normal and negate the hyperthyroid implications of the high FT_4 value. Dexameth-

TABLE 4.7. Patterns of apparently conflicting thyroid function test results, diagnostic considerations, and methods to exclude hyperthyroidism.

Patterns of test results				
STSH	FT_4	FT_3	Diagnostic Considerations	Exclusion of hyperthyroidism
N or ↓	↑	↓	5′-Deiodinase deficiency Euthyroid sick syndrome Dexamethasone Propranolol	A normal STSH excludes hyperthyroidism, an ↑ in STSH level after TRH negates the hyperthyroid implications of a subnormal STSH. The low FT_3 makes hyperthyroidism very unlikely.
N or ↑	↑	↓	Cholecystographic media Amiodarone Euthyroid sick syndrome	The N or STSH excludes hyperthyroidism.
N or ↑	↑	↑	Peripheral resistance to thyroid hormone Amphetamines Inappropriate excess pituitary TSH secretion	The N or ↑ STSH excludes hyperthyroidism, excepting that caused by inappropriate pituitary TSH secretion. A pituitary workup may be needed.
↓	↑	N	T_4 toxicosis Patient taking T_4 overdose Acute psychosis Hyperemesis gravidarum Heparin	A low RAIU suggests T_4 overdosage. If needed, a T_3 suppression test may be helpful in psychotic patients or those on heparin. Hyperemesis patients can be observed for spontaneous recovery.
↓	N or ↓	↑	T_3 toxicosis Patient taking T_3	A normal FT_4 and a normal to elevated RAIU suggest T_3 toxicosis. A low FT_4 and a low RAIU suggest that the patient is taking T_3.

asone and propranolol depress 5′-deiodinase in peripheral tissues, but not in the pituitary. Propranolol elevates the FT_4, but only in a dose of 160 mg per day or higher.[38] Cholecystographic media [39] and amiodarone [40] diminish 5′-deiodinase activity in the pituitary and in peripheral tissues. Reduced conversion of T_4 to T_3 in the pituitary is perceived as a state of thyroid hormone deprivation, and TSH is released. The STSH value may be normal or increased, but the subnormal value of hyperthyroidism is not seen unless there is hyperthyroidism or one of the nonhyperthyroid causes for a low STSH (e.g., hypercortisolism, psychiatric depression, inanition, or acute illness). The low FT_3 immediately impugns the hyperthyroid implication of a high FT_4 value. T_4 toxicosis (i.e., hyperthyroidism where only the serum T_4 concentration is elevated) is well recognized,[2,41,42] but the FT_3 is normal to high-normal, not low. Also T_4 toxicosis is primarily a disease of older patients.[2,41,42] Of course, amiodarone, an iodine-rich compound, can induce hyperthyroidism. In this case, both FT_4 and FT_3 levels will be elevated, and the STSH will be suppressed.

Euthyroid Sick Syndromes

Abnormal in vitro test results are common in euthyroid patients hospitalized with nonthyroidal illnesses.[1,32-36] Low T_3 and FT_3 values are uniformly present. T_4 values may be high, normal, or low.[35] FT_4 values are high or normal, but not low.[36] STSH values can be high, normal, or low. Thus, it may become necessary to exclude hyperthyroidism or hypothyroidism, depending upon the pattern of tests results obtained.

A subnormal STSH with an elevated FT_4 might suggest hyperthyroidism. A subnormal FT_3 value is evidence against that diagnosis. If the patient were hyperthyroid, the FT_3 value, if not elevated, should be high normal or at least well within normal limits. Furthermore, TRH would produce an increment in the STSH value that, even if not as great as ordinarily seen in euthyroid patients, is adequate to exclude hyperthyroidism, where no response would be expected. If the FT_3 value and the TRH test are inconclusive, and hyperthyroidism remains a viable consideration, it may be useful to fall back on the old T_3 suppression test.[43] Since there is no fall in the RAIU in hyperthyroidism after T_3 administration (25 μg three times daily for 4 to 5 days), any substantial reduction in the RAIU would exclude hyperthyroidism. If the administration of T_3 is deemed too hazardous in a sick or older patient, and hyperthyroidism as a complicating condition seems reasonably likely, it may be prudent to administer antithyroid drugs promptly. The added burden of hyperthyroidism to another major illness can be life threatening, and treatment should not be delayed. The antithyroid drugs can be discontinued after recovery and the patient reevaluated for hyperthyroidism.

An elevated STSH might suggest hypothyroidism, especially in conjunction with a low T_4 value. However, the FT_4 value would be normal, and thus provide strong evidence against more than minimal hypothyroidism. When the diagnosis of hypothyroidism is not clear-cut by clinical and laboratory evaluation, there is seldom any urgency to start treatment. The elevated STSH value in euthyroid sick patients returns to normal within a few weeks along with recovery from the basic illness.

Miscellaneous Nonthyroidal Conditions

Peripheral resistance to thyroid hormone may be sporadic[44] or an autosomal dominant familial trait.[45] Resistance usually involves all body tissues. There is goiter, elevated FT_4 and FT_3 values, without hyperthyroidism. The serum STSH is normal or elevated, findings incompatible with hyperthyroidism. The administration of additional T_4 or T_3 will suppress STSH, but will also increase the blood level of the specific thyroid hormone administered.

If resistance to thyroid hormone is limited to the pituitary, the pituitary will respond as in hypothyroidism with an increased release of TSH. The output of thyroid hormone is increased. Since peripheral tissues are not resistant, hyperthyroidism develops.[46] This is an analogous situation to patients with inappropriate pituitary oversecretion of TSH (neoplastic or dysfunctional).[47] Serum STSH levels are elevated, in spite of supernormal serum concentrations of the thyroid hormones. The pituitary is routinely enlarged on CT imaging. Other evidence of a pituitary tumor may be present.

If resistance is limited to peripheral tissues, and the pituitary is not involved, the pituitary release of TSH will not increase in response to the peripheral deficiency of thyroid hormone activity. These patients are hypothyroid in spite of normal serum T_4, FT_4, and STSH levels. Even TRH tests give normal data. It will be necessary to induce elevated FT_4 levels to restore these patients to the euthyroid state. Since the diagnosis is dependent upon recognition of clinical findings of hypothyroidism in spite of "normal" thyroid function tests, it is probable that patients with this form of hypothyroidism are overlooked. The magnitude of this problem in unknown.[46] If this disease is suspected, one might search for nonspecific electrocardiographic findings of hypothyroidism, including flat or inverted T waves, low amplitude P, and QRS waves, and sinus bradycardia. Also, there may be a shortened preejection period for any given heart rate; mild anemia, especially macrocytic; elevated cholesterol and triglyceride levels; and elevated creatine phosphokinase and lactic dehydrogenase levels. If the patient has clinical findings of hypothyroidism, and several of these abnormalities, treatment with thyroid hormone may be prudent. Reversal of laboratory abnormalities and clinical improvement would support the diagnosis of peripheral resistance to thyroid hormone.

In acute psychiatric illness[48] and hyperemesis gravidarum,[49] there may be temporary and spontaneously reversible FT_4 elevations. Also, there may be a low STSH value and a subnormal (but not a flat) TRH response, probably because of poor nutrition. FT_3 levels may be normal or low if there is poor nutrition. The possibility of T_4 toxicosis is ruled out by the response of the TSH to TRH. Diagnosis of acute psychiatric illness and hyperemesis gravidarum are not difficult. For the psychiatric patient, a T_3 suppression test could be done if necessary. Hyperemesis gravidarum fortunately is of short duration. The clinical evaluation may make hyperthyroidism unlikely. Even if the clinical picture is somewhat suspicious (e.g., there are goiter and other findings suggestive of hyperthyroidism), it might still be better to wait 1 or 2 months until the hyperemesis subsides, rather than starting treatment with antithyroid drugs. The mild hyperthyroidism that might be hard to diagnose will be well tolerated, whereas the administration of antithyroid drugs during pregnancy has well-known risks.[50]

Patients with heterophilic antibodies to rabbit immunoglobulins can have spurious elevations in conventional TSH assays.[51] The FT_4 value is normal. If a TRH test were done, the increment in TSH level would be normal, in spite of the elevated baseline TSH level. Adding fresh rabbit serum before the addition of the first antibody normalizes the assay. Monoclonal STSH assays partially eliminate this problem. However, an occasional patient with antibodies to mouse immunoglobulins will have spuriously high STSH values. Addition of mouse serum will normalize the STSH.

Thyroid hormone administration alters thyroid function tests in predictable fashion. Pure T_4 products may elevate FT_4 values and suppress STSH levels, even in doses that need not produce clinical features of hyperthyroidism. Whether these are slight overdoses or not may be argued. Some advocate that as long as FT_3 levels are normal, no reduction in T_4 dosage is needed.[52,53] Since it is easy to maintain patients on doses of levothyroxine that do not elevate FT_4 values, and also maintain normal FT_3 and STSH values, there is little reason not to do so.

Administration of T_3 to euthyroid people, whether factitious or by prescription, suppresses FT_4 and STSH values, and elevates FT_3 values. Although T_3 is seldom used for long-term treatment, factitious self-administration of T_3 does occur. In T_3 toxicosis, FT_4 values are normal to high-normal, and RAIU values are elevated or at least "normal." In T_3 overdosage, FT_4 and RAIU values are very low. Factitious self-administration of thyroid hormone is more common in members of the health profession or their relatives.

Dextrothyroxine has been used for hyperlipidemia. Since it is much less potent than the levo form, larger doses have been given. Since radioimmunoassay methods do not distinguish between the stereoisomers, high FT_4 values are obtained.[54] There are also increased FT_3 values because of the higher concentration of the dextrothyroxine precursor for T_3. This is not ordinarily a clinical problem because dextrothyroxine is used infrequently and is an improbable choice for factitious thyrotoxicosis.

The high FT_4 and FT_3 values seen with high doses of amphetamines are compatible with hyperthyroidism, a diagnosis that might be suggested by the nervousness, tremor, and tachycardia these people manifest.[55] However, the STSH level is elevated, excluding hyperthyroidism.

Heparin-treated patients may have elevated FT_4, normal FT_3, and normal or low STSH values.[56] The elevated FT_4 may be an in vitro artifact related to the generation of free fatty acids.[57] The exclusion of T_4 toxicosis[41,42] depends upon the clinical evaluation, and possibly a TRH test. On rare occasions one might consider a T_3 suppression test. The administration of heparin is not likely to be factitious, hence awareness of the effect of heparin in thyroid function tests should prevent confusion.

Cost-Effective Clinical Laboratory Diagnosis of Thyroid Dysfunction

In clinical practice the physician wants to know whether his or her patient has hypothyroidism, hyperthyroidism, or neither. The specific cause of misleading abnormal thyroid function test data is less important. Indeed, if the explanation is not obvious or simply determined, usually it may safely be ignored, as long as thyroid dysfunction or serious illness has been excluded.

The primacy of the clinical evaluation in the establishment of a working diagnosis of thyroid dysfunction is axiomatic. However, much of the literature dealing with misleading thyroid function test data does so with little regard to the clinical findings. Consequently, one might get the impression that the potential for diagnostic error and mismanagement is very great. This need not be so. Most of the time, misleading laboratory data are obtained in patients without clinical evidence of thyroid dysfunction, especially without goiter. Sometimes the cause of the abnormal laboratory data is obvious (e.g., acute psychiatric patients, hyperemesis gravidarum, or drugs known to influence in vitro thyroid function tests). Furthermore, the clinician who suspects hypothyroidism, while laboratory tests suggest hyperthyroidism, or vice versa, surely would reconsider before instituting therapy.

There is no need to restate the clinical features of hypothyroidism and hyperthyroidism. These entities are well described in standard medical texts, and are familiar to all physicians. Nevertheless, there are guidelines that will help physicians to maintain a proper focus as they attempt to establish or exclude working diagnoses of hypothyroidism or hyperthyroidism.

In this further discussion it is assumed that the FT_4 method is the Gammacoat two-step procedure, that the FT_3 assay includes the use of polyethylene glycol to remove T_3 binding antibodies, and that a STSH assay is used.

General Principles Relative to the Diagnosis of Thyroid Dysfunction

Thyroid dysfunction affects the body generally, and both symptoms and physical findings are therefore multiple. Single complaints, even if consistent with thyroid dysfunction (e.g., dry skin, fatigue, constipation, hair loss, nervousness, tremor, tachycardia, or weight loss), are seldom caused by thyroid dysfunction. Not only are symptoms of thyroid dysfunction multiple, but they are also concordant (i.e., favoring consistently either hypothyroidism or hyperthyroidism). Discordant symptoms make thyroid dysfunction less likely. Similarly, physical findings should be consistent with the diagnosis suggested by the history. The laboratory data will then serve to confirm or exclude the working clinical diagnosis. Since more than one test of thyroid function is generally employed, consistent patterns of results are of most diagnostic value. Isolated abnormalities are less meaningful, and discordant values provide evidence against thyroid dysfunction.

Clinical Laboratory Diagnosis of Hypothyroidism

Since considerable attention has already been paid to the euthyroid sick syndromes, the following discussion concentrates on the ambulatory patients with whom most practitioners must be prepared to deal. Figure 4.1 presents a flowsheet for the laboratory confirmation of the clinical diagnosis of hypothyroidism. The end point in each instance is an etiologic diagnosis or group of diagnoses that may or may not require precise differentiation.

An important contributing factor to the erroneous diagnosis of hypothyroidism is the failure to establish an etiology for the disease. Common to nearly all patients who are unnecessarily treated for hypothyroidism is the absence of any credible etiology. Since the treatment of primary hypothyroidism is the same, regardless of the cause, knowledge of the etiology does not change the treatment, but is of fundamental importance in confirming the need for treatment.

Hypothyroidism may be goitrous or nongoitrous (Fig. 4.1). In goitrous hypothyroidism the problem is defective or inefficient synthesis and secretion of thyroid hormone, so that the thyroid gland undergoes compensatory hyperplasia in response to increased TSH release by the pituitary. By far the most common cause of goitrous hypothyroidism in the United

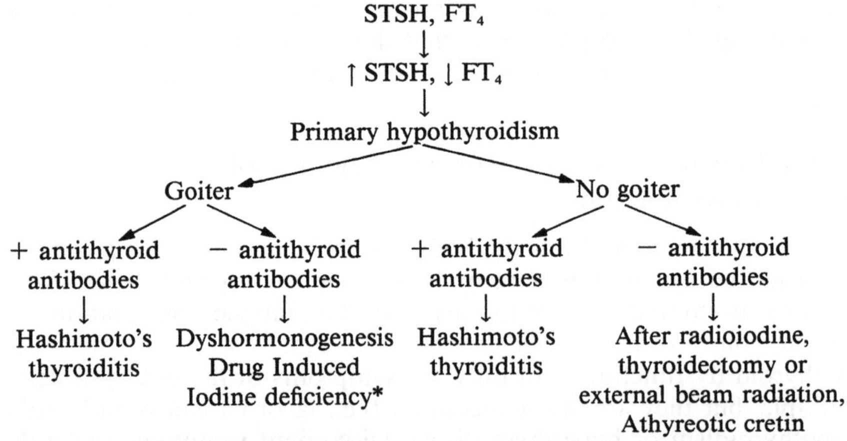

FIGURE 4.1. Laboratory confirmation of clinical diagnoses of hypothyroidism. *Almost nonexistent in the United States.

States is Hashimoto's thyroiditis. A firm, lobulated or diffuse goiter in a patient with clinical features of hypothyroidism (often mild) should immediately suggest the need for obtaining a microsomal antibody titer to confirm the diagnosis of Hashimoto's thyroiditis.[58] The antithyroglobulin assay, flocculation method, is less sensitive. The antithyroglobulin antibody by radioimmunoassay (RIA) may be more sensitive. Five percent of patients tested for possible Hashimoto's thyroiditis in my laboratory had elevated antithyroglobulin antibody titers by RIA, without positive microsomal antibody titers. Other causes of goitrous hypothyroidism are much less common, but include the enzyme deficiencies of dyshormonogenesis, and drug-induced goiter (most commonly iodide or lithium). Drugs tend to produce goitrous hypothyroidism in conjunction with some other basic thyroidal defect (e.g., Hashimoto's thyroiditis or dyshormonogenesis). Iodine deficiency is the most common cause of goitrous hypothyroidism in endemic areas worldwide, but it is virtually nonexistent where iodine prophylaxis is practiced.

When the FT_4 is low and the STSH is elevated, a diagnosis of primary hypothyroidism and the indications for treatment with levothyroxine are clear. Some patients, most commonly those with goiters of Hashimoto's thyroiditis, may have normal FT_4 values, but elevated STSH values. Many thyroidologists would also treat these patients with levothyroxine. The earliest evidence of failing thyroid function is an accentuated response to TRH of a normal or high-normal baseline STSH value. The necessity for treatment of these patients with levothyroxine is controversial.

Nongoitrous hypothyroidism is most commonly iatrogenic after radioactive iodine therapy or thyroidectomy, and on rare occasions after intensive radiation therapy to the neck (Fig. 4.1). The advanced stage of the atrophic form of Hashimoto's thyroiditis also may be nongoitrous. Nongoitrous hypothyroidism is nearly always severe, and thus seldom a diagnostic problem. Much less common forms of nongoitrous hypothyroidism are congenital athyreosis and secondary hypothyroidism. The appropriate diagnoses are suggested by the cretinoid appearance of the former, or the evidence of other target organ deficiencies in pituitary disease.

If there is nothing favoring any of the above etiologies, a diagnosis of hypothyroidism should be made with great reluctance, and only after compelling and comprehensive laboratory confirmation. Positive antithyroid antibody titers confirm the diagnosis of Hashimoto's thyroiditis. Negative titers do not exclude that diagnosis. About 15% of patients with Hashimoto's thyroiditis do not have positive antibody tests. Lack of antithyroid antibodies is more common in younger patients. Although one could confirm the diagnosis with a needle biopsy, this is not ordinarily necessary if the evidence for hypothyroidism is clear. Needle biopsy is better reserved for those with discrete nodules, large hypofunctional defects on thyroid imaging, or goiters that either fail to regress or enlarge in spite of treatment with thyroxine.[59]

The diagnosis of hypothyroidism is usually simple, and involves only limited testing. Note that the FT_3 test has no value in the confirmation of hypothyroidism because it is frequently normal, even when the patient is hypothyroid clinically and has low FT_4 and elevated STSH levels. The RAIU has no place in the diagnosis of hypothyroidism. Thyroid imaging would be useful only for the evaluation of a possible associated neoplastic thyroid process.

Clinical Laboratory Diagnosis of Hyperthyroidism

Figures 4.2 to 4.5 present flowsheets for the laboratory confirmation of clinical diagnoses of hyperthyroidism. An elevation in the FT_4 with a subnormal STSH is virtually diagnostic of hyperthyroidism (Fig. 4.2). One could conceive of possible exceptions (e.g., a patient on amphetamines or with peripheral resistance to thyroid hormone who might at the same time be so sick or malnourished that the STSH value is low). However, it is unlikely that such patients would also have the goiters and high RAIU values of hyperthyroidism. With typical test data, a RAIU value that is "normal" or elevated suggests thyroidal hypersecretion as the cause of the hyperthyroidism.

The differentiation of Graves disease from toxic multinodular goiter and the toxic autonomous nodule is simply accomplished by physical examination, supplemented by thyroid imaging if desired. A very low RAIU necessitates study for less common causes of hyperthyroidism (Table 4.5). The presence of goiter would make subacute thyroiditis or Jodbasedow disease prime considerations (Fig. 4.3).

STSH, FT_4, FT_3*

↓

↓ STSH, ↑ FT_4, ↑ FT_3

↓

Hyperthyroidism

↓

RAIU

↓

N or ↑ RAIU

↙	↓	↘
Diffuse goiter	Multinodular goiter	Large single nodule
Graves disease	Toxic multinodular goiter	Toxic autonomous nodule

FIGURE 4.2. Laboratory confirmation of clinical diagnoses of hyperthyroidism. *FT_3 assays are optional, and primarily useful when the clinical diagnosis is possible or probable, but not certain.

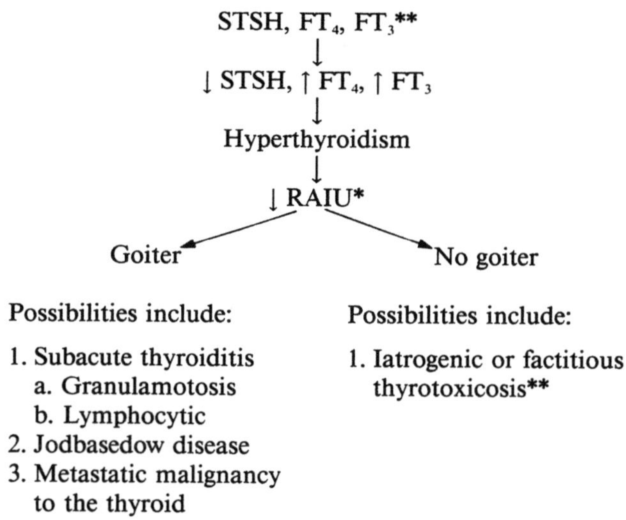

FIGURE 4.3. Laboratory confirmation of clinical diagnoses of hyperthyroidism. *See Table 4.5 for uncommon causes of hyperthyroidism with low RAIU. **FT_3 may be normal or elevated depending upon the type of thyroid hormone the patient is taking.

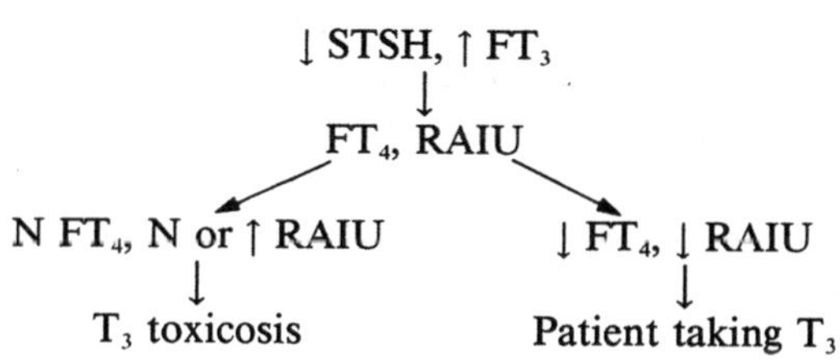

FIGURE 4.4. Laboratory confirmation of clinical diagnoses of hyperthyroidism.

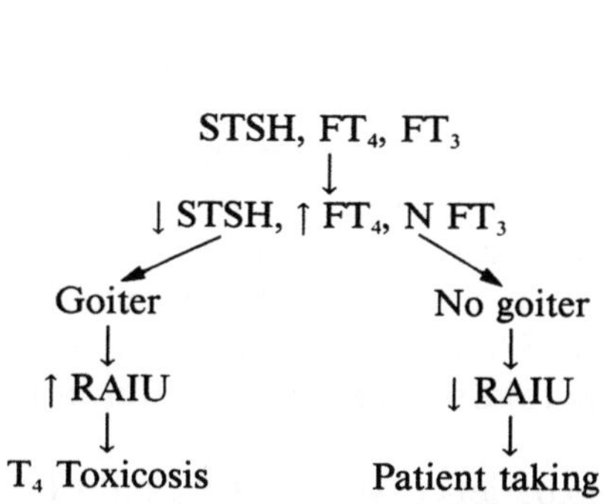

FIGURE 4.5. Laboratory confirmation of clinical diagnoses of hyperthyroidism.

Metastatic malignancy to the thyroid on rare occasions may cause hyperthyroidism. The absence of goiter would suggest ectopic sources of thyroid hormone (toxic struma ovarii, or massive metastases of thyroid carcinoma), iatrogenic or factitious overdosage of thyroxine, triiodothyronine, or dextrothyroxine (Fig. 4.4 and 4.5).

An elevated (or even a nonsuppressed) STSH value in a patient with clinical hyperthyroidism and elevated FT_4 and FT_3 values suggests inappropriate pituitary TSH secretion, neoplastic or nonneoplastic (Table 4.7). The difficulty of differentiating this entity from the uncommon syndrome of pituitary resistance to thyroid hormone has already been discussed.

Summary

Avoiding pitfalls in thyroid function testing requires a working diagnosis of hypothyroidism or hyperthyroidism based on clinical findings. A history of drug ingestion may explain misleading laboratory data. The implications of nonthyroidal illness must be considered.

The STSH test is the keystone for confirmation and exclusion of thyroid dysfunction. FT_4 and FT_3 assays are useful adjuncts. The free thyroxine index is less reliable. T_4 and T_3 tests measure inactive protein-bound hormone, and are not reliable thyroid function tests. TRH tests may help after inconclusive or inconsistent baseline function tests. The RAIU is not a primary test of thyroid function, but is important to differentiate between hyperthyroid syndromes with normal or elevated values and those with low values. The T_3 suppression test may help differentiate between hyperthyroidism and nonthyroidal causes of abnormal in vitro thyroid function tests.

Discussion by Dr. Leonard Wartofsky

You indicate that TBPA exclusively binds T_4. But TBPA can bind T_3 as well, depending upon the buffer system employed.

You say that the unbound hormone fraction correlates well with thyroid function; but the correlation is with "metabolic state."

In regard to your tabulation, I believe that Table 4.2 is well conceived and could be useful. However, the usefulness of any thyroid function test is relative to the diagnosis being considered. Hence, the T_3RIA is of no value in the diagnosis of hypothyroidism, the RAI uptake is important for the diagnosis of thyroiditis, the FTI is useful in nonthyroidal illness, and so forth.

I have no personal experience with the FT_3, but wonder about its utility in the diagnosis of hypothyroidism.

You indicate that the TBG concentration is increased in lymphosarcoma. Could you provide a reference for that observation? Also, you are not clear on whether increased or decreased TBPA means concentration or binding affinity. Especially in the euthyroid sick syndrome, it is believed to be binding affinity that is influenced, presumably by some circulating inhibitor.

You have listed amphetamines and heparin (Table 4.6) as drugs causing abnormal thyroid function tests for reasons "other than abnormal protein binding." I believe there are data that indicate at least some of the aberrations in thyroid function tests may be due to binding changes.

In discussing function test abnormalities in the euthyroid sick syndrome you say you would not treat patients with elevated STSH levels who have normal FT_4 values. Why would you not recommend T_4 therapy here on an empiric basis, as the corollary to your earlier recommendation of antithyroid drug therapy for questionable diagnoses of hyperthyroidism?

In your discussion of tissue resistance to thyroid hormones, with resistance limited to peripheral tissues, does the hypothyroidism imply coexistent Hashimoto's or some other cause of reduced thyroid reserve? If not, why shouldn't a normal gland compensate?

You say that the STSH is elevated in patients receiving high doses of amphetamines, and cite the work of Hershman's group published in 1981.[55] However, they could not have been using a STSH assay at that time.

You say, with regard to patients with Hashimoto's thyroiditis, goiter, and elevated STSH but normal FT_4 values, "necessity for treatment of these patients with levothyroxine is controversial." I would ask what arguments may be put forth in favor of *not* treating?

In Figure 4.3, the double asterisk, referring to patients with iatrogenic or factitious hyperthyroidism, says that FT_3 may be normal or elevated depending upon the type of thyroid hormone the patient is taking. I would wonder whether free T_3 would be elevated irrespective of the kind of thyroid hormone taken?

Response by Dr. Joel I. Hamburger

Dr. Wartofsky has raised a number of important issues that need amplification.

I said TBPA "almost" exclusively binds T_4, referring to the in vivo situation. Griffin[1] said very little T_3 is bound to TBPA.

Dr. Wartofsky is correct. The unbound fraction of thyroid hormone may have no correlation with thyroid function (e.g., in factitious thyrotoxicosis, when thyroid gland function is suppressed, FT_4, FT_3, or both are elevated). However, the suggested "metabolic status" seems somewhat imprecise. Perhaps we could agree that free hormone levels correlate more reliably than total hormone levels with thyroid hormone activity in tissues.

Dr. Wartofsky is correct that the usefulness of any given thyroid function test depends upon the use to which it is put. In Table 4.2 the tests were classified in terms of usefulness for the evaluation of any thyroid function problem. The specific uses of tests that are still useful are discussed subsequently.

The FT_3 assay is as useless in the diagnosis of hypothyroidism as is the total T_3 assay.

An increased TBG capacity in lymphosarcoma was reported by Borst and Burman.[2] Dr. Wartofsky is correct that I have not been careful enough to distinguish binding protein concentration from binding protein affinity. Of course, the impact on assays for total T_4 or T_3 concentration would be the same whether protein binding concentration or affinity were increased or decreased. Hence the difference is only of theoretical importance.

Borst and Burman[2] said, "amphetamine induced hyperthyroxinemia is mediated through a direct central stimulation of the thyroid axis or by induction of a state of peripheral thyroid hormone resistance." Morley et al[55] said the same thing. With regard to the high FT_4 caused by heparin, the latest information I have (presented at the 1987 Washington, DC meeting of the American Thyroid Association) is that it is an in vitro artifact caused by generation of free fatty acids.[57]

Dr. Wartofsky asks why I would not advise levothyroxine treatment for patients with nonthyroidal illness, elevated STSH values, but normal FT_4 values, especially since I said that I would probably favor antithyroid drugs for patients with nonthyroidal illness and subnormal STSH values, elevated FT_4 values but normal or even low FT_3 values. Clearly my comments on these issues need amplification. First, what is it that I would not treat?; and second, why not? I would not treat a patient with nonthyroidal illness whose only indication of primary hypothyroidism were a mild to moderate elevation in the STSH. If that finding were indicative of primary hypothyroidism there ought to be other clinical laboratory findings pointing to hypothyroidism and a reasonable etiology for the presumed disease. For example, a history of radioactive iodine therapy for hyperthyroidism, a thyroidectomy scar, or a firm lobulated goiter consistent with Hashimoto's thyroiditis, along with positive antithyroid antibody assays, would be findings suggesting a high probability of thyroid disease. The presence of clinical findings of hypothyroidism, including infraorbital edema, hung-up deep tendon reflexes, coarse skin, and so forth would be persuasive, but when present would be associated with more overt elevations of the STSH, and a subnormal FT_4 value. Thus, if the only abnormality is a mild to moderate elevation in the STSH value, I wouldn't treat the patient with thyroxine. Why not? Because the patient probably doesn't have hypothyroidism. It is well known that elevated STSH levels are common in nonthyroidal illness, and they may be spontaneously reversible with recovery from nonthyroidal illness.[60–62] At worst the patient might have mild hypothyroidism, and that would pose no special threat to recovery from the nonthyroidal illness, after which reevaluation for hypothyroidism should be simpler. Indeed, some say that these modest increments in STSH (or TSH) values may be beneficial for patients with nonthyroidal illness.[63] On the other hand, I would not argue with a physician who elected to give 25 to 50 μg of levothyroxine to such a patient. The chances of doing harm would be exceedingly small. In contrast, why would I advise antithyroid drugs to a patient with nonthyroidal illness and subnormal STSH, elevated FT_4, but a low or low-normal FT_3 value? Actually, I said that "it may be prudent" to do so. I was referring to the patient who is too sick to tolerate a T_3 suppression test of the RAIU. Although I didn't say so, I was referring to a patient who had an enlarged thyroid gland. I was speaking with appreciation that severely ill patients have usually experienced weight loss and often have tachycardia and sweating (fever and infection). Since even mild hyperthyroidism can add substantially and immediately to the debility of severe nonthyroidal illness, I would give antithyroid drugs even if the diagnosis of hyperthyroidism were not certain.

Why shouldn't the normal thyroid gland compensate for resistance to thyroid hormone limited to the periphery? Because the pituitary doesn't tell it that it should. Since resistance to thyroid hormone is only in peripheral tissues, the pituitary is satisfied with usual levels of thyroid hormone. Thus there is no increase in TSH release. Levels of thyroid hormones in peripheral tissues are "normal," but peripheral resistance leaves those tissues insufficiently supplied and therefore hypothyroid.

Dr. Wartofsky is right again. Hershman's group was not using a STSH assay in 1981 when he found high TSH values in patients on amphetamines. But STSH assays have no special advantage over conventional TSH assays in identifying elevated serum TSH concentrations. STSH assays are superior to conventional TSH assays only in differentiating subnormal from normal serum TSH concentrations.

Why not order levothyroxine for patients with goiters (i.e., smaller goiters), antibody evidence for Hashimoto's thyroiditis, elevated STSH assays, but normal FT_4 levels? I frequently treat these patients both to prevent further thyroid enlargement and to assure that adequate supplies of thyroid hormone will continue to be available. However, it has been said that "The augmented TSH levels will further increase iodine trapping and, if there is adequate functional thyroid reserve, sufficient organification and hormone synthesis may ensue to return T_4 and T_3 levels toward normal and the patient to a state of compensated hypothyroidism."[64] This is a concept that has been reiterated by other authorities.[65,66] Appreciation that an elevated TSH may be found in 5% of the population,[67] and 13% of healthy elderly individuals[68] gives some perspective to the implications of universal treatment for these people. Most of them do not experience further deterioration in thyroid function, at least not within a few years.[68–70] The recommendation to treat those with higher TSH levels, higher antithyroid antibodies, and symptoms of hypothyroidism, while observing the rest[67] makes sense to me.

Can one have iatrogenic-factitious thyrotoxicosis in LT_4-treated patients without an elevation in the FT_3? This is one of the most controversial issues in clinical thyroidology today. It deals with how we define the limits of normality in terms of thyroid function tests; and this is the sine qua non for their rational use in clinical practice. In this case the issue is what is the most sensitive laboratory finding(s) of LT_4 overdosage?

For many years elevated T_4 values were considered acceptable in patients who appeared to be clinically euthyroid while being treated with daily LT_4 doses of 200 to 400 μg.[71] In 1974, Stock et al.,[72] giving graded increments of LT_4, showed that LT_4 doses of 100 to 200 μg per day were adequate to restore elevated TSH levels to the normal range. On those doses 96% of patients had serum T_4 levels within normal limits, although the mean value was slightly higher than for euthyroid controls. Serum T_3 levels were also normal, but the mean was slightly less than for euthyroid controls. At that time TSH assays were not sensitive enough to distinguish between normal and subnormal. Those authors concluded that the higher mean T_4 value did not mean LT_4 overdosage because the mean T_3 value was reduced. They suggested that the serum T_3 assay was the test of choice to evaluate LT_4-treated patients. This idea was consistent with the concept of T_4 as an inactive prohormone from which most of the active hormone, T_3, in the serum and intracellular sites was derived by monodeiodination.

The 1982 reformulation of Synthroid that increased tablet strength led Fish et al[73] to reevaluate replacement doses of LT_4 in hypothyroidism. As in the earlier study of Stock et al, the last named (and probably the senior) investigator was Jack Oppenheimer. Once again graded doses of LT_4 were given until elevated STSH levels were restored to normal. The mean daily replacement dose of new Synthroid was only 119 ± 19 μg, compared to 169 ± 66 μg with old Synthroid. However, using new Synthroid normalization of STSH seemed to require LT_4 doses that produced a significantly elevated mean T_4 value (rather than the high-normal mean T_4 value with the old Synthroid). The mean T_3 value was the same as that for controls (rather than lower than controls as with old Synthroid), and the control value for the serum T_3 was substantially higher in this study, even though the same method was used in both studies. Finally, the conversion of T_4 to T_3 was substantially less for patients in the second study. Since mean values for both T_4 and T_3 were higher in the second study, the conclusion seems unavoidable that the actual (as opposed to nominal) LT_4 doses given were higher. Fish et al offer no explanation for these differences. They were content with the observation that their findings were similar to those of Hennessey et al,[74] who determined that an average daily LT_4 dose of 127 ± 39 μg or 1.72 μg/kg corrected hypothyroidism without abolishing the TSH response to TRH.

Since LT_4-treated hypothyroid patients lack the 18% of circulating T_3 that normally comes from the thyroid, they require a higher concentration of the T_4 substrate for T_3 production to maintain the T_3 levels considered critical to regulate TSH.[62,72] Hence, Fish et al. concluded that T_3 may be more important than T_4 in regulating serum TSH levels.

For comparison, a group of my patients with hypothyroidism after ^{131}I therapy (32 patients) or with Hashimoto's thyroiditis with small to impalpable thyroid glands (20 patients) were euthyroid with normal STSH values on daily LT_4 doses averaging 110 ± 33 μg or 1.5 μg/kg. Although these doses were only modestly lower than those in the two previous studies, Fish et al[73] emphasize that rather small increments or decrements in FT_3 concentrations have important effects on stimulation and suppression of TSH by the pituitary. Be that as it may, 51 of my 52 patients had normal FT_4 values, and the one elevated value was only 2.3 ng/dl (NR = 0.6–2.2). The mean FT_4 value was 1.42 compared to 1.1 for controls. The mean FT_3 value was 2.56 pg/dl (NR = 2–6), somewhat less than the 3.5 mean value for controls. These data show that it is not necessary to have elevated FT_4 values to restore the euthyroid state clinically and in terms of a normal STSH value, just as Oppenheimer's group indicated in their 1974 report.

In contrast, 11 of my recent patients on LT_4 with undetectable TSH by the STSH assay had an elevated mean FT_4 value of 2.73 ng/dl, and a mean FT_3 value of 3.78 pg/dl that was correspondingly higher than that for the above patients with normal STSH values, even though all FT_3 values were within the normal range. Although these patients had no convincing clinical findings of thyrotoxicosis, their LT_4 doses were higher than necessary to normalize STSH values. Did they have iatrogenic thyrotoxicosis? Some say that as long as serum T_3 levels are not elevated, high FT_4 and subnormal STSH values are not important.[73,75] Does this make sense? Do normal FT_3 values assure euthyroidism? Certainly not! Indeed, Fish et al found difficulty reconciling the high serum T_3 values in iodine-deficient areas with the idea that T_3 played the primary role in pituitary TSH regulation. They speculated that the pituitary threshold for TSH suppression is reset in such patients. Is it also reset in the many patients with symptomatic hypothyroidism, subnormal FT_4, elevated STSH, and normal FT_3 values?[76] Is it reset in the opposite direction in T_4 toxicosis where FT_4 values are elevated, STSH values are undetectable, but FT_3 values are within normal limits?[77] In all these situations it is the STSH that correlates better than the FT_3 with intracellular thyroid hormone activity.

Fish et al[73] reject the applicability to humans of Larsen's studies in rats showing increased pituitary T_4 to T_3 conversion compared to peripheral tissues, studies suggesting a dominant role for T_4 in the regulation of pituitary TSH secretion.[78] Nevertheless, adherence to the idea that one can infer the level of intracellular thyroid hormone activity most reliably from serum T_3 levels requires circumlocutions too delicate for easy acceptability.

The newer idea that the STSH level is the best indicator of tissue thyroid hormone activity[79,80] permits more uniformly consistent correlations between clinical and laboratory findings. STSH levels are always elevated in overt hypothyroidism, even if FT_3 levels are confusingly "normal." STSH levels are always suppressed in hyperthyroidism, regardless of FT_4 levels (that may be normal in T_3 toxicosis) or FT_3 levels (normal in T_4 toxicosis). The only problems are that STSH levels may be elevated before there is convincing clinical evidence of hypothyroidism (discussed above in response to Dr. Wartofsky's question), and STSH levels may be undetectable without obvious clinical evidence of hyperthyroidism (e.g., in some patients with euthyroid Graves disease, multinodular goiter, and most particularly in those treated with LT_4 doses that may be excessive).

Are the higher LT_4 doses that suppress, rather than just normalize, STSH values harmless, as some maintain[71,73,75,81] or do they have the potential for causing damage? Hyperthyroidism clearly causes loss of bone mineral density. Coindre et al[82] showed a loss of bone mineral density in response to treatment with a thyroid hormone product containing T_3 (20 μg) and T_4 (100 μg) daily. Since they did not monitor patients with a STSH assay, they could not exclude unrecognized overdosage. In any event, a well-controlled follow-up study from another center showed no adverse impact of even mildly supraphysiologic doses of LT_4 on bone mineral density.[83]

Markers for metabolic activity in various tissues are altered in hyperthyroidism, and concordant albeit less severe abnormalities have been found in LT_4-treated patients with high FT_4, subnormal STSH, but normal T_3 values.[29-31] Perhaps the findings that might cause most concern relate to the heart and the liver. Reduced systolic time intervals were reported in LT_4-treated patients with high mean FT_4 and normal mean FT_3 levels.[84] Those findings were subsequently challenged on the basis of inconsistency and inadequate controls.[85] However, another report showed an increased mean nocturnal heart rate in LT_4-treated patients who had TSH levels unresponsive to TRH.[86] An elevated Wayne index score provided clinical evidence for thyrotoxicosis for 10 of 25 patients with elevated FT_4 and normal T_3 levels during LT_4 treatment.[87] Suggestive of subclinical liver damage from inappropriately high LT_4 doses is the finding of an increase in glutathione *S*-transferase levels.[88]

If hyperthyroidism, like hypothyroidism, is a continuum with the most severe cases at one end, and the least severe at the other, as Toft suggests,[89] it should be anticipated that thyrotoxicosis in those with lesser degrees of LT_4 overdosage may be hard to prove by even sophisticated test methods. What, then, should the prudent physician do in ordinary clinical practice, where judgments must be made on the basis of routinely available clinical assays? Should he or she follow the advice of those who say that without compelling data suggesting that slight overreplacement with LT_4 is harmful one can disregard laboratory findings that suggest it,[74] (e.g., a subnormal STSH)? Or should he or she follow the principle that current tests of thyroid function are more reliable than clinical assessment?[89] In my opinion, the burden of proof lies with those who advise ignoring the LT_4 overdose implications of a subnormal STSH. Unless they can provide compelling data suggesting the need for (of course excepting treatment of thyroid cancer, when full suppression of TSH may be desirable) and long-term safety of overdosage, it might be preferable to prescribe the lower doses of LT_4 that produce clinical euthyroidism and normal STSH values as well.

Therefore, the answer to the basic question of whether LT_4-induced thyrotoxicosis is always associated with elevated FT_3 levels depends upon whether one is inclined to accept or reject the evidence that there may be subclinical thyrotoxicosis without an elevated FT_3.

Commentary by Dr. Robert Volpé

My, how this controversy does go on! I will indeed be pleased to respond to the point that you have raised, namely does thyroxine overdosage-induced thyrotoxicosis occur without an elevated serum T_3 (or free T_3) level?

The use of L-thyroxine in therapeutics is generally for two objectives. The first of these is to suppress thyroid tissue for whatever reason, for example, goiter suppression, or following primary treatment for thyroid carcinoma, or secondly for the treatment of hypothyroidism. These objectives are clearly different since in the former, TSH should be suppressed, while in the latter precise replacement consonant with optimal good health is the final objective.

We will deal with hypothyroidism initially. It is indeed clear that the best current arbiter of normal replacement of thyroxine in hypothyroidism would be a TSH value (performed by a "sensitive" assay) that is in the middle of the normal range. That would mean that the patient is completely finely tuned as far as it is humanly possible to accomplish this. The big question is, is there any harm to a patient when the TSH is suppressed, providing that the total serum triiodothyronine (or free T_3) is in the middle of the normal range? That is, does a somewhat elevated serum thyroxine have deleterious implications, in the face of a completely normal serum T_3 level? My current view is that it does not, and that it is not worth the manipulation and individual attention to thyroxine dosage that some have claimed are necessary for "fine tuning." In my view, serum levels of triiodothyronine correlate better with clinical status than do the serum thyroxine levels. This may be made evident by assessing the clinical severity of hyperthyroidism in patients with subacute thyroiditis versus those with Graves disease in relation to their levels of serum thyroxine and triiodothyronine. Quite often the levels of serum thyroxine are similar in the two groups of patients, yet those with Graves disease are obviously much more ill. This seems to clearly correlate with the disproportionately higher levels of triiodothyronine in those patients. Ingbar and his colleagues[90] have indicated that the triiodothyronine levels correlate better than serum thyroxine levels with clinical status. It should be mentioned that most patients with hypothyroidism initially have at least a little remaining intrinsic thyroid function (greater than zero up to virtually normal thyroid function). Thus one would expect that a full physiologic replacement dose of thyroxine would suppress the thyrotropin response to thyrotropin-releasing hormone (TRH) and would also suppress TSH. Only in those patients who have absolutely no thyroid function initially, would a physiologic replacement dose of thyroxine be unable to blunt the thyrotropin response to TRH or suppress TSH. That is, if before treatment there was some degree of feedback, however minimal (with some thyrotropin suppression, however minimal), then clearly less than a physiologic replacement dose would be required to bring that particular person to a precisely "normal" status where either thyrotropin response to TRH was perfectly normal and the sensitive TSH assay was in the middle of the normal range.[91]

In all other instances, a blunted response of thyrotropin to TRH or a suppressed "sensitive" TSH value would not necessarily mean that the patient is receiving too much thyroxine. It is the same situation as would be expected if a person who had normal thyroid function was treated with even a small dose of thyroxine. Because the thyroid function was initially normal, even a small dose of thyroxine would be sufficient to blunt the thyrotropin response to TRH or suppress the sensitive TSH value. Such a person would not be overreplaced until the dosage exceeded the reduced thyroidal secretion of hormone induced by the suppression of thyrotropin. This point has indeed been made earlier by Stock et al[72] in 1974 and again by a colleague and me[92] in 1981. In the latter publication we argued that the total serum triiodothyronine was the best test for monitoring L-thyroxine therapy.[92] The next point has to do with possible ill effects from dosages of thyroxine that will suppress TSH. I have argued above that one would not obtain an overreplacement dosage merely by suppressing TSH. Thus an elevated T_4 with a normal serum T_3, and a suppressed TSH value is not an indicator of overdosage, under these circumstances.

Bantle[93] has pointed out once again what the significance of normal serum T_3 is under the circumstances of exogenous T_4 administration. He has stated that the supranormal levels of serum T_4 in levothyroxine-replaced hypothyroid patients may be necessary to achieve normal circulating T_3 concentrations. It has been calculated that approximately 18% of the total extrathyroidal pool of T_3 is normally derived from direct thyroidal secretion, whereas the remaining 82% originates from the peripheral conversion of T_4 to T_3. Since in hypothyroid patients, thyroidal secretion of T_3 is reduced or absent, nearly all circulating T_3 must be derived from peripheral monodeiodination of T_4, and higher than normal serum T_4 concentrations are thus necessary to maintain normal serum T_3 concentrations. Practicing physicians should therefore not be surprised by moderately elevated T_4 concentrations in levothyroxine-replaced patients and should not interpret such values as indicative of hyperthyroidism.

There is certainly a controversy about bone density. Ross and his colleagues[94] have suggested that dosages of thyroxine that suppress TSH were associated with osteopenia, at least as determined by direct photoabsorptionometry when taken at the junction of proximal two thirds and distal one third of the nondominant radius. However, this had not been confirmed by Ahmann and his colleagues.[83] The metabolic disturbances as described elsewhere in this volume by Hamburger in terms of systolic time intervals and increase in glutathione *S*-transferase levels have either been disputed on grounds of inadequate controls or do not appear to have any clinical correlation.

To me as a clinician, the matter is also one of long experience. We have recently reviewed 100 patients who have been on thyroxine 0.2 mg per day for 20 years following removal of the thyroid for thyroid carcinoma. These patients virtually all have suppressed TSH determinations. We have compared them with a random control group matched for age and sex but otherwise healthy. The comparisons had to do with clinical evidence of either bone disease or heart disease. There was no difference in these areas between these two groups, nor in their sense of well-being (unpublished data). From observations like this, it is difficult for me to agree that "fine tuning" is really necessary. It is also true that there are patients who do have complaints when they feel they are being overmedicated. What is required to determine whether they are truly complaints secondary to the medication itself, or whether the complaints are functional is to study this question with an appropriate control group. This control group should also be taking some placebo and the exhortations must be entirely similar. Such a study is yet to be done. Until such a study is complete and indicates with certainty that the medication is causing clinical symptoms, I remain unmoved. "Fine tuning" is, for me, logistically enabling, expensive, and in the final analysis, probably unappealing.

Commentary to Dr. P. Reed Larsen

While we have not done a systematic study such as that reported by Fish et al[73] and therefore our conclusions are tentative, our experience using sensitive TSH assays and T_4 replacement is not similar to that described in their report but much more consistent with reports of Gow et al.[88] For patients with primary hypothyroidism (most of whom have serologic evidence of Hashimoto's thyroiditis), we give approximately 0.8 μg of either the reformulated Synthroid or Levothroid per pound. Such patients have TSH values in the 0.1 to 0.15 range (Allegro assay, normal 0.5 to 5.0), serum T_3s 100 to 120 (normal mean 120), and serum T_4 levels ~8 to 11. The dual regulation of TSH by both T_3 and T_4 in the rat is, as you indicate, now well established. I don't think that we are as yet clear on whether the independent regulation by locally produced T_3 occurs in the hypothalamus or the pituitary, but probably in both, given the recent data on the influence of thyroid hormone on TRH synthesis. From an historical point of view, it is interesting to note that the hypothesis regarding an exquisitely maintained serum T_3 concentration as the TSH regulator raised by Fish et al in their article is virtually identical to the one we initially raised 10 years earlier as an explanation for observations in the iodine-deficient rat.[95] The latter study demonstrated that when rats are given an iodine-deficient diet, there is a rapid increase in serum TSH as T_4 falls despite the fact that the serum T_3 does not change significantly. While the possibility of direct T_4 feedback on the hypothalamic-pituitary axis was considered to be probable (local T_3 production not as yet having been recognized), we also postulated that it could be argued that TSH secretion is finely modulated in such a way as to maintain a constant serum T_3 level. We speculated that a small initial drop in serum T_3, below the discrimination level of the T_3RIA could have been the stimulus to the initial rise in serum TSH. The identical hypothesis for the observations in humans with mild hypothyroidism was mentioned by Fish et al. While this hypothesis may be correct (in both rat and man), it is unfortunately untestable since the change in T_3 necessary to initiate the process is, by definition, unmeasurable.

Some of the best, albeit indirect, evidence regarding the importance of local T_4 to T_3 conversion in regulating human TSH in man comes from Dr. Braverman's laboratory.[39] In a small but carefully controlled study they showed that 25 μg of T_3 per day for 2 days (in divided doses) caused a reduction in TRH-induced TSH release of 50%. When T_3 administration was repeated in the same individuals after iopanoic acid, the maximum TSH release after TSH was increased to about twice that found with T_3 administration alone. This suggested that there was a significant decrease in feedback inhibition of TSH release due to the iopanoic acid-induced blockade of T_4 to T_3 conversion. This small study is not definitive, however, since the mean T_3 levels were 193 in the controls whereas those in the iopanoic plus T_3 patients were 137 ng/dl and this could have influenced the responses.

I agree that to normalize serum T_3 values in patients with no thyroidal T_3 secretion, one will have to increase the T_4 production by a factor that would allow normal T_3 production exclusively from deiodination. This could be predicted to be an increase in serum T_4 of about 2 μg/100 ml. It may require even more given Nicoloff's data that there is saturation of a low K_m enzymatic process for deiodination of T_4 to T_3 as serum T_4 increases.

Finally, one does not have to cite iodine deficiency studies to convince oneself that serum T_4 is somehow important in the regulation of TSH secretion in man. The data of Bigos et al[96] collated in my review in the *New England Journal of Medicine*[3] shows clearly that when serum T_4 is reduced, even with a somewhat elevated serum T_3 concentration, basal TSH and TRH-induced TSH release are enhanced. This agrees so well with all reports on this subject that it seems difficult to escape the conclusion that man and rat are similar in this regard.

Hopefully, careful studies of the effects of T_3 versus T_4 replacement correlating serum T_3 values with TSH measurements by a sensitive assay will be forthcoming soon so that this issue can be settled. In the meantime, my clinical practice is to use the T_4 dose of 0.8 μg/lb body weight as a rough approximation for replacement. This usually results in serum free T_4 index values that are about 9 to 11, serum free T_3 indices of 100 to 120 (slightly subnormal), and serum TSH values that are low-normal. If serum TSH falls below 0.5 μg/ml (Allegro assay), then I will reduce the T_4 dose slightly except in patients in whom TSH suppression is the goal of therapy.

Commentary by Dr. Anthony D. Toft

Before the development of radioimmunoassays for thyrotropin (TSH) in the early 1970s, the recommended daily dose of thyroxine for patients with primary hypothyroidism was 200 to 400 μg.[53,97] The appropriate dose was assessed mainly on clinical grounds and, if serum protein-bound iodine or total thyroxine (TT_4) was measured, it was customary to regard values that were in the upper part of the normal range or even elevated as acceptable, as this would compensate for the loss of thyroidal secretion of triiodothyronine (T_3). Using TSH assays, however, it became clear that a dose of between 100 and 200 μg of thyroxine daily was sufficient to restore elevated levels in primary hypothyroidism to normal in some 90% of patients.[72] The practical roles of the TSH assay in this group of patients were to confirm adequacy of thyroxine dosage and patient compliance. For example, it became well recognized that patients tended to become more diligent tablet takers in the few days prior to a clinic visit, resulting in the seemingly anomalous combination of raised TT_4 and TSH. As the radioimmunoassays for TSH were not sufficiently sensitive to differentiated suppressed (thyrotoxic) from normal (euthyroid) levels, basal TSH measurements could not be used to assess whether thyroxine dosage was excessive and the decision to reduce therapy depended upon clinical assessment. Although it had been established that many seemingly euthyroid patients receiving 100 to 200 μg of thyroxine daily showed a lack of TSH response following intravenous thyrotropin-releasing hormone

(TRH),[98] this finding was taken to reflect an increased sensitivity of the pituitary thyrotroph to minor changes in thyroid hormone levels[99] that would not be recognized by other target organs. Is the thyrotroph the most sensitive target tissue for thyroid hormones?

In the rat, at least, the anterior pituitary is different from other target organs in that not only are nuclear T_3 receptors 78% occupied but also approximately half of the T_3 is derived from intrapituitary monodeiodination of T_4. In contrast, liver and kidney nuclear T_3 receptors are only 50% occupied and less than 20% of nuclear T_3 is derived from T_4,[3] the majority coming from T_3 in the circulation. If the same mechanism were present in man, thyrotroph suppression would not necessarily be accompanied by changes in other target organs and subclinical hyperthyroidism* induced by treatment with thyroxine would be of no clinical significance.

There is, however, increasing evidence that doses of thyroxine sufficient to suppress thyrotroph secretion cause tissue hyperthyroidism in other organs such as heart, kidney, bone, and liver despite the lack of clinical evidence of thyrotoxicosis.

Red Cell Sodium. In retrospect, the first indication that the standard replacement therapy of the time was excessive was made by Goolden et al[100] who demonstrated that not only were red cell sodium levels raised in hyperthyroidism but also, to a lesser extent, in clinically euthyroid patients with thyroid carcinoma taking 300 μg of thyroxine daily. Although serum TSH levels were not measured, it is almost certain that thyrotroph function was suppressed in these patients. The increased red cell sodium in hyperthyroidism is due to a reduced number of pump sites in the erythrocyte membrane. Wilcox and Levin[101] were able to demonstrate a significant decrease in these sites, by means of measuring ouabain binding capacity, in thyroxine-treated patients the majority of whom were receiving 150 to 200 μg daily. It is unfortunate that almost 50% of the clinically euthyroid patients had raised FT_3 levels, although there was no correlation between FT_3 and ouabain binding capacity.

Heart Rate and Myocardial Contractility. Tachycardia is a feature of hyperthyroidism and at one time it was customary to record sleeping pulse rate to distinguish thyrotoxic patients from those with anxiety. A similar increase in nocturnal heart rate has been demonstrated in subclinical hyperthyroidism (Fig. 4.6). In this study, heart rate was continuously monitored during a 24-hour period of normal activities in healthy volunteers before and after taking thyroxine for a period of 24 days in a dose sufficient to abolish serum TSH response to TRH. TT_4 levels rose within the normal range from 93 to 128 nmol/L (7.4 to 10.4 μg/dl) and

*In this context, subclinical hyperthyroidism is defined as normal or raised TT_4 or FT_4 normal TT_3 or FT_3 and absent TSH response to TRH or undetectable basal sensitive TSH (STSH) in an asymptomatic patient.

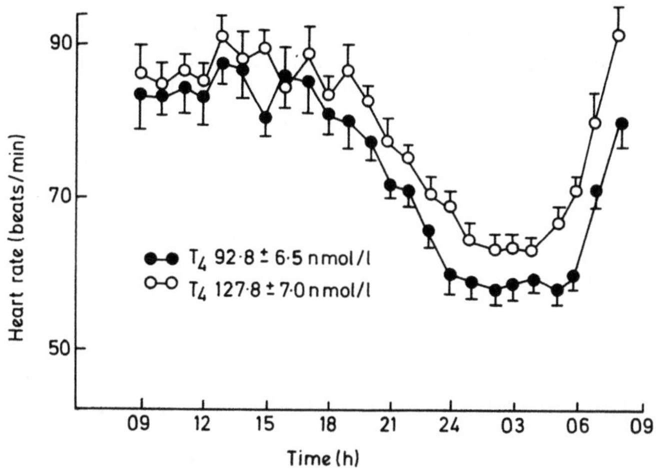

FIGURE 4.6. Mean heart rate (beats/min ± SE calculated hourly for 25 hours in subjects before (●) and during the 24th day of T_4 administration (○). (Reprinted from Bell et al[86] with the permission of *Clinical Endocrinology.*)

TT_3 remained unchanged. Although the mean increase in nocturnal heart rate was only 7 beats per minute, this was the first demonstration that the myocardium might be as sensitive as the thyrotroph to minor changes in thyroid hormone levels within the normal range.[86]

Jennings et al[84] showed increased myocardial contractility, based on shortened systolic time intervals (STI), in a series of patients taking 100 to 200 μg of thyroxine daily. Although serum TT_4 levels were elevated and TSH response to TRH absent in the majority, TT_3 levels were normal and the patients were clinically euthyroid. In some, reduction of thyroxine dose was accompanied by restoration of STI to normal.

Urinary Sodium Excretion. Peripheral oedema may be a feature of hyperthyroidism and can be partly explained by daytime sodium and water retention.[102] In the same subjects described in the heart rate study,[86] it was shown that although 24-hour sodium excretion and urine flow were unchanged by thyroxine administration, there was a significant decrease in the day-to-night ratio of these indices, similar to that which occurs in hyperthyroid patients. Although it is likely that such changes are due to a direct effect of thyroid hormones on renal tubular cell metabolism, it is possible that the phenomenon may be secondary to the increased nocturnal heart rate and the consequential increases in cardiac output, renal blood flow, and glomerular filtration rate.

Liver. Before effective treatment of thyrotoxicosis was available, hepatobiliary complications were well recognized. Liver biopsy often showed morphologic changes, including glycogen depletion, fatty change, and cirrhosis. These were not considered to be due to a direct effect of thyroid hormones,[103] but the combined result of several insults such as cardiac failure, infection, hypoxia, and malnutrition. Although severe and prolonged hyperthyroidism is now fortunately rare, minor hepatic abnormalities can be demonstrated in some thyrotoxic patients. Electron microscopy of liver biopsy samples shows nonspecific changes in the hepatic organelles.[104] Biochemical evidence of hepatic dysfunction depends upon the marker used, but in one series abnormal sulphobromophthalein retention was reported in 8%, but raised bilirubin and alanine aminotransferase less frequently.[105] If, however, an extremely sensitive marker of hepatocellular damage, such as glutathione *S*-transferase (GST) is measured in hyperthyroid patients, raised levels are recorded in the majority.[106] There is little evidence to suggest that hypothyroidism affects liver function, but these patients will be treated with thyroxine and as a result many will have raised TT_4 or FT_4 normal TT_3 or T_3 and suppressed TSH. Do such patients develop hyperthyroidism of the liver? An additional risk must be the high concentrations of thyroxine presented to the liver, through the portal vein, following an oral dose.

As shown in Figure 4.7, GST levels became abnormal as a result of thyroxine treatment in four of eight patients presenting with primary hypothyroidism. In three of the four, abnormal GST was associated with a raised FT_4. These results have been confirmed in a much larger series.[88]

Bone. There are now two reports of reduced bone density in premenopausal women receiving thyroxine replacement therapy. In both, the mean daily dose of thyroxine was 170 μg, resulting in raised serum T_4, normal

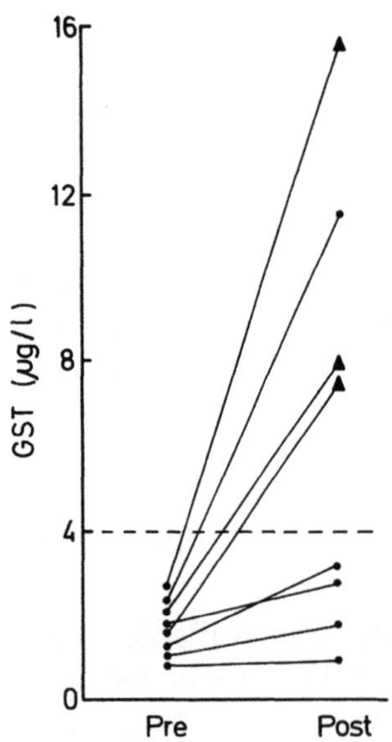

FIGURE 4.7 Serum glutathione *S*-transferase measurements in patients with primary hypothyroidism before and after thyroxine therapy. ▲, Patients with raised FT_4 concentrations after treatment. All patients had normal TT_3 and FT_3 values after treatment. (Reprinted from Beckett et al[106] with the permission of the *British Medical Journal*.)

T_3, and, in the majority, suppressed TSH. Ross et al[94] showed a reduction of 4% and 9% in the nondominant wrist of patients treated for greater than 5 and 10 years, respectively (Fig. 4.8). Paul et al,[107] using a dual photon technique, reported a reduction in bone density of 18% in the femoral neck in patients treated for longer than 5 years, but no change in the lumbar vertebrae. The effect of long-term thyroxine therapy on postmenopausal osteoporosis, although potentially serious, is unknown.

How Should the Correct Dose of Thyroxine Be Assessed—Biochemically? Given the above evidence that doses of thyroxine sufficient to suppress thyrotroph secretion are associated with hyperthyroidism in other tissues, it would seem logical to adjust replacement therapy to maintain serum TSH in the normal range. The limitations on sensitivity of the TSH radioimmunoassays meant that in practice it was necessary to demonstrate a normal TSH response to TRH to exclude thyrotroph suppression. This test was not only time consuming for a busy clinic but also not without adverse effects[108,109] and did not find favor for this particular purpose. The recently developed sensitive assays for TSH (STSH), however, which detect levels of less than 0.1 mU/L and consistently distinguish euthyroid from hyperthyroid sera, have rendered the TRH test

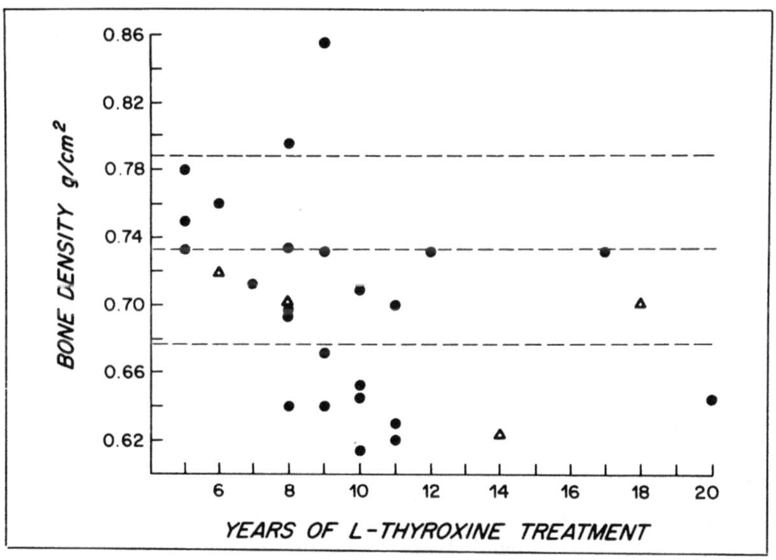

FIGURE 4.8. Bone density in 28 premenopausal women taking L-thyroxine for 5 or more years. The normal mean ± SD for bone density in premenopausal women is indicated by the broken lines. ●, patients with absent or subnormal TSH response to TRH; △, patients with normal TRH test result. (Reprinted from Ross et al[94] with the permission of *The American Journal of Medicine.*)

redundant.[110,111] An accurate assessment of thyrotroph function can now be made from a basal TSH measurement.

Gow et al[88] measured STSH in 104 clinically euthyroid patients who had been taking thyroxine replacement therapy for a mean period of over 3 years and in whom a constant dose had been prescribed for at least 3 months. Patients were subdivided into three categories on the basis of STSH level and their details and the results of free and total thyroid hormone levels are shown in Table 4.8. Whereas FT_4 was elevated in 86% of patients with undetectable STSH, FT_3 and TT_3 were normal in 67% and 98%, respectively.

Serum markers of peripheral tissue response to thyroid status were also measured in the three groups. These included alanine aminotransferase (ALT), GST, γ-glutamyltransferase (GGT), creatine kinase (CK), thyroxine binding globulin, and sex-hormone-binding globulin. The results are shown in Figure 4.9. Significantly higher ALT, GST, and GGT and lower CK were found in those patients with undetectable STSH when compared to patients with normal values, demonstrating a clear relationship between tissue thyroid status and thyrotroph function. In contrast, there was no relationship between serum markers and T_3 measurements and many patients with normal T_3 had evidence of tissue hyperthyroidism. It would appear from this study that measurement of serum T_3, although

TABLE 4.8. Comparison of patient details and thyroid function tests subdivided according to TSH concentrations in 104 patients taking T_4 replacement therapy.

	Mean (SD or range)		
	Group A (low TSH) (n = 37)	Group B (normal TSH) (n = 42)	Group C (high TSH) (n = 25)
Dose (μg)	162 (60)	127 (43)	116 (47)
Age (yr)	53 (13)	53 (14)	49 (14)
Wt (kg)	66 (14)	69 (13)	72 (12)
Duration (yr)	3.2 (0.3–15)	3.8 (0.3–22)	3.7 (0.5–21)
TSH (μU/ml)	<0.1†	1.4 (0.13–5.9)	24.3 (6.4–103)†
Free T_4 (ng/dl)	2.8 (1.4)†	1.7 (0.3)	1.2 (0.4)§
Free T_3 (ng/dl)	0.5 (0.2)†	0.4 (0.1)	0.3 (0.1)‡
T_4 (μg/dl)	11.9 (3.5)†	9.5 (1.6)	7.4 (2.3)†
T_3 (ng/dl)	126 (28)†	105 (17.5)	96 (23.1)

Reference ranges: TSH, 0.14–5.9 μU/ml; free T_4, 0.8–1.8 ng/dl; free T_3, 0.28–0.55 ng/dl; T_4, 5.6–12.0 μg/dl; T_3, 77–182 ng/dl. Reprinted from Gow et al[88]: Relationship between pituitary and other target organ responsiveness in hypothyroid patients receiving thyroxine replacement. J Clin Endocrinol Metab 1987;64:364. © by The Endocrine Society.

*$p < .005$ versus group B.

†$p < .001$ versus group B.

‡$p < .05$ versus group B.

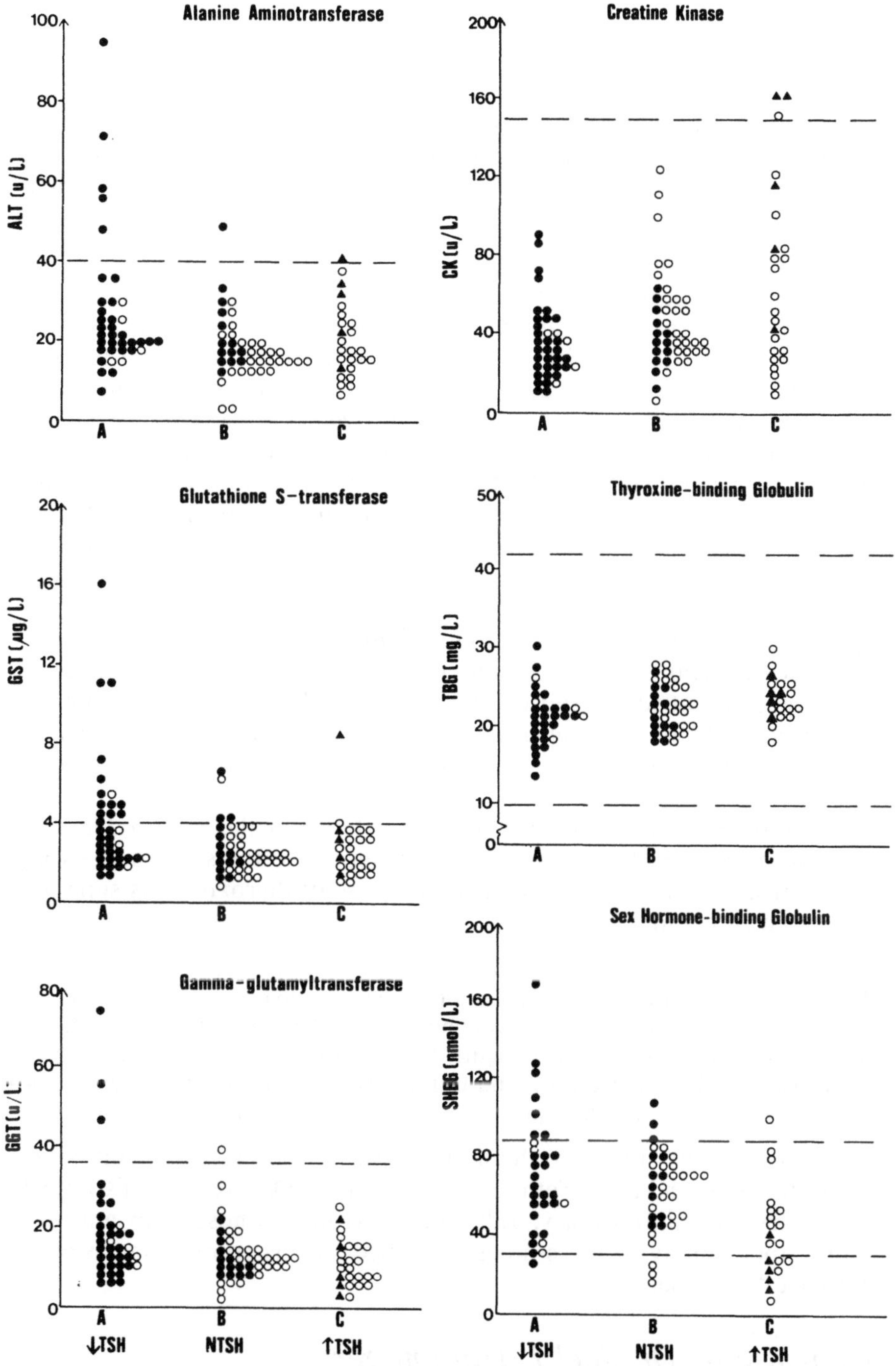

FIGURE 4.9. Peripheral tissue markers in serum from 104 patients taking a fixed T_4 dose grouped according to STSH concentrations (A, undetectable; B, normal; C, raised). Upper reference range values (ALT, GST, GGT, and CK) and reference ranges (TBG and SHBG) are indicated by the broken lines. Values for patients with high (●), normal (○), and low (△) FT_4 concentrations are indicated. (Reprinted from Gow et al[88]: Relationship between pituitary and other target organ responsiveness in hyperthyroid patients receiving thyroxine replacement. J Clin Endocrinol Metab 1987;64:364. © by The Endocrine Society.

it correlates well with clinical assessment of thyroid status,[75,112] can no longer be regarded as the measurement of choice for assessing thyroxine dosage as it is not an indication of thyroid status at a tissue level, pituitary or otherwise.

—or Clinically? There are some who would question the relevance of these minor abnormalities of target organ function in patients receiving thyroxine replacement therapy who are asymptomatic,[73,75] and there are others who claim that clinical judgment is better than any test of thyroid function.[113] After all, what evidence is there of increased morbidity or mortality in patients treated with 200 or even 300 μg daily? However should the onus of proof that slightly too much thyroxine is harmless not rest with the disbelievers? Most physicians will now accept that the aim of replacement therapy should be to maintain a normal STSH, and if this is undetectable in patients with primary (not secondary) hypothyroidism, measurement of FT_4 or TT_4 will provide an indication of the magnitude of reduction in dosage necessary.

There have been significant developments in thyroid function testing in the last 20 years and it would seem sensible to utilize these, however sensitive, and not to depend on the fickle variable of clinical judgment that, in current laboratory terminology, lacks not only sensitivity but also specificity—at least in the diagnosis of minor degrees of thyroid dysfunction.

I should like to end by quoting from a letter in the *British Medical Journal* of October 25, 1986 written in response to the study suggesting that clinical judgment of the appropriate dose of thyroxine was superior to thyroid function tests:[113]

> By showing a pronounced discrepancy between clinical assessment based on the Wayne Index (which was not intended for patients on thyroxine therapy) and the results of thyroid function tests, they conclude that the tests must be at fault. As an analogy to this flawed logic, imagine if diabetes mellitus were still diagnosed by tasting the urine for sweetness and this procedure were used as the gold standard against which a new test, plasma glucose concentration, was judged.

Acknowledgments. I wish to thank my close colleagues, Drs. Geoff Beckett, Sadie Gow, and John Seth, of the University Department of Clinical Chemistry, Royal Infirmary, Edinburgh, without whose help and encouragement many of the studies described in this commentary would not have taken place.

Response by Dr. Joel I. Hamburger

Just as the editorial work in these proceedings was coming to a conclusion, an article in *JAMA* from Braverman's group[114] provided further evidence that T_4 doses that increase the serum T_4 concentrations above the normal range and produce subnormal TSH levels, without elevating serum T_3

levels, can cause loss of bone mineral. An accompanying editorial[115] advised maintaining STSH levels within normal limits when giving T_4 to correct hypothyroidism. As Dr. Volpé said: "My, how this controversy does go on!" I will add that it will not go away.

Discussion by Dr. James C. Sisson

To place great confidence in the direct assays of FT_4 and FT_3 seems premature. Indeed, all estimates of free hormone values suffer imprecision in the presence of nonthyroidal illness, and the direct assays of free thyroxine concentrations have not been shown to be more reliable than those derived from the FTI.[116] I cannot agree with the statement that calls the T_3RU and FTI obsolete tests. Some physicians, like myself, prefer to know both the total hormone and the free hormone levels that give perspective to the patient's status; this is especially important when the total T_4 (and total T_3) value is low and the free T_3) level is likely to be less reliable. Moreover, much of the wisdom in the practice of medicine is derived from experience, and a change to a marginally more accurate test sacrifices the difficult-to-replace experience obtained with the previous test.

It is not clear that STSH assays will regularly give results that can be equated with an absent response of TSH to injected TRH. Some physicians prefer to treat patients with thyroid cancer to the lowest measurable level of TSH secretion, a state that may not be consistently revealed by the STSH.

In patients who have nonthyroidal illness and are also suspected of having hyperthyroidism, the T_3 suppression test may be helpful in establishing a diagnosis. However, many patients with nonthyroidal illness are investigated by many modes, including radiologic studies that use contrast media. Diagnostic procedures that employ iodine-containing agents will vitiate the uptake measurements in the suppression tests.

One gets the idea that FT_4 and FT_3 values should be used to periodically evaluate patients taking thyroxine. Why not use the STSH alone for such laboratory assessment of such patients? If the clinical suspicion is primary hypothyroidism, as is often derived from the presence of goiter or the history of a thyroidectomy or radioiodine treatment, why should both STSH and T_4 measurements be necessary as implied in Figure 4.1? The STSH (or TSH) should suffice in these circumstances.

Response by Dr. Joel I. Hamburger

My experience with the Clinical Assays two-step FT_4 assay and the Amersham FT_3 assay extends over 2 years in thousands of patients. The Clinical Assays FT_4 assay is the only FT_4 kit method that is reliable in familial dysalbuminemic hyperthyroxinemia where the FTI gives falsely elevated values. It is because these and similar patients are seen with increasing frequency that I investigated FT assays, and found, as Dr. Braverman and colleagues[15] said, that the Clinical Assays test works. I have no personal experience with the euthyroid sick syndrome, because I deal exclusively with ambulatory patients. However, it has been reported by others that the FTI is not reliable in these patients.[33] The FT_3 assay is an important improvement over T_3RIA assays that measure the total serum T_3 concentration, and is altered by situations in which T_3 binding is abnormal. Attempts to correct the T_3RIA by the T_3RU is less effective, and more expensive. I, too, like to know the total T_4 concentration because this provides information on protein binding. The T_4RIA assay is a more direct and reliable way of obtaining this information than the T_3RU.

I recognize that it is easier for one who operates a small private lab that only performs eight assays than for a large institutional lab to change methods. However, the only way to acquire the experience is to try newer assays.

Dr. Sisson is correct that subnormal STSH values do not preclude responsiveness to TRH. However, undetectable STSH values nearly always mean that there will be no response to TRH.

I do not use the FT_3 to monitor patients treated with thyroxine, for reasons exhaustively discussed already. I agree that the STSH alone may be all that is necessary. The FT_4 offers a measure of insurance, but may be redundant.

I think we are approaching the position that Dr. Sisson suggests, that is, full reliance on the STSH. However, most of us feel more confident when important clinical judgments are verified by more than a single assay.

Discussion by Dr. Ian D. Hay

In commenting on Dr. Hamburger's contribution on cost-effective thyroid function testing, I find few areas to criticize. At Mayo, we too have been unimpressed by the clinical performance of the analog FT_4 methods but, to date, have had only limited experience with the Amersham FT_3 assay.

I totally agree that the STSH assay is now the "keystone" of thyroid function testing, and also believe that currently there is "little reason not to" titrate T_4 doses to permit the maintenance of serum STSH levels within the normal (euthyroid) range. It is certainly worth emphasizing that a subnormal STSH value is not diagnostic of hyperthyroidism, but rather has the same implication as an impaired TSH response to TRH.

Response by Dr. Joel I. Hamburger

We are now evaluating a free T_3 assay kit by Diagnostic Products Corporation that seems to give data as reliable as the Amersham kit, but is simpler and less expensive.

Discussion by Dr. John E. Freitas

I believe that the article by Mendel et al confirms that the elevated FT_4 is usually an in vitro artifact related to the heparin-induced generation of free fatty acids.[117] This work also indicates that T_4-binding globulin in patients with the euthyroid sick syndrome is dysfunctional and that the low plasma T_4 concentrations and high percent free T_4 values seen in such patients is not due to an inhibitor of T_4 binding.

Discussion by Dr. John T. Dunn

Since your emphasis is on a cost-effective strategy, it would help for you to give the actual costs of the various tests. Related to this, I agree that the STSH is by far the most valuable test for general thyroid function, because it is a bioassay. It is more expensive than most other tests, and less valuable in certain situations, such as following the patient on antithyroid drugs for Graves disease. For the latter, an initial T_4 and T_3 resin uptake, and subsequent T_4s during follow-up are usually quite satisfactory. At least in the labs I am familiar with, the T_4 and T_3 resin uptake are a great deal cheaper than the free T_4 and free T_3. Perhaps that will change with increased use. Please comment on the *relative* costs of these.

You bring up a common problem, that of the patient with multinodular goiter, a suppressed TSH, and seeming clinical euthyroidism. It has been occasionally suggested that this situation may occur because of a change in the pituitary setpoint. Another interpretation, one that I favor, is that the pituitary is smarter than we are in diagnosing hyperthyroidism, and that these patients are in fact subtly hyperthyroid, and should probably be treated. I would be interested in your comments.

We all have problems in drawing diagrams, because there will inevitably be exceptions. For example, if a patient comes in with obvious clinical hyperthyroidism, a diffuse goiter, and exophthalmos, is it necessary to do a TSH?

I would certainly favor treating patients with elevated TSH's but normal T_4s as hypothyroid. Why not treat them when they have minimal hypothyroidism rather than waiting for it to be gross? It reminds me of the arguments of 25 years ago, when many clinicians insisted on gross myxedema before accepting a diagnosis of hypothyroidism.

I was interested in the comment that metastatic malignancy to the thyroid can cause hyperthyroidism. For my personal interest, can you give me more details on this?

Jodbasedow disease certainly causes hyperthyroidism, and we see it in ares of endemic goiter. I have not seen good data on the RAIU in this situation, but would speculate that if the iodine is enough to make them overproduce hormone, the uptake would probably not be suppressed. Do you have specific information on this point?

Response by Dr. Joel I. Hamburger

Cost is an important consideration in the use of thyroid function tests. Costs vary widely in my locality, depending upon whether the work is done in hospitals, commercial labs, or physicians' labs. For example, in one leading community teaching hospital in metropolitan Detroit the charge for a T_4 assay is \$64. More reasonable comparisons of costs may be provided by looking at Blue Shield reimbursement for the various procedures, since most physicians and commercial labs accept Blue Shield payments and Blue Shield is by far the largest single health insurance provider in Michigan. Table 4.9 presents current Michigan Blue Shield payments for representative tests.

I, too, have some concern for seemingly euthyroid patients with multinodular goiters and suppressed STSH values. They may well have impending thyrotoxicosis. If FT_4 and FT_3 are high-normal, and the patients are older than 50, I would advise prophylactic therapy with ^{131}I. However, frequently they are younger, and FT_4 and FT_3 values are only at mid-normal levels. I observe these patients, because I am not sure I can trust the pituitary to predict future events reliably.

For a patient with obvious clinical hyperthyroidism, a diffuse goiter, and exophthalmos, one might ask if it is necessary to do any tests. At least theoretically this will be Graves disease 95+% of the time. One could give a dose of ^{131}I based on the size of the goiter, between 8 and 18 mCi, and more than 90% of the patients would be cured and ready to start replacement thyroxine in 6 to 12 weeks. They could then be treated with 0.1 mg daily of levothyroxine and monitored by clinical findings. Once again, more than 90% of those patients would do well. Testing could be reserved for the remaining few who needed it. This system would probably be the ultimate in cost effectiveness. If physicians were protected from any liability for the occasional errors that would occur, it would be practical and produce enormous savings.

Unfortunately, from the cost perspective, we physicians are driven by tradition, training, and the legal profession to perform complete workups, and to fully document with objective data every step we take. Since laboratory tests are subject to technical errors, single tests are not considered

TABLE 4.9. Michigan Blue Shield payments for selected thyroid function tests.

Test	Blue Shield payment
T_4	\$ 8.00
T_3 resin uptake	8.00
FT_4	18.75
FT_3	18.75
STSH	20.00

adequate documentation, and therefore, batteries of tests are done. Patients must pay for this until their elected representatives change the system.

I, too, favor treating patients with elevated STSH levels, even though FT_4 values are within normal limits. If a single STSH value is only slightly elevated, I would use a TRH test to demonstrate the augmented response that would support a diagnosis of thyroid functional impairment. Since the treatment would nearly always have to be for life, I like to have more substantial evidence that it is necessary than just a single slight elevation in the STSH value. How much further I would pursue the workup would depend upon the clinical setting, and especially the evidence for an etiology for the presumed thyroid disease. For example, if the patient had the typical goiter and antibody findings of Hashimoto's thyroiditis, a surgical scar, or a history of ^{131}I therapy for hyperthyroidism, obviously I would settle for limited laboratory documentation of the need for thyroxine therapy. If there were no obvious etiology, I would be inclined to advise more extensive study or observation prior to committing the patient to lifelong treatment.

Metastatic malignancy to the thyroid gland can cause disruption of thyroid follicles and discharge of thyroid hormone and a form of spontaneously resolving hyperthyroidism with low RAIU. The mechanism is similar to that in subacute thyroiditis. It has been reported with pancreatic carcinoma,[118] lymphoma,[119] and, I believe, breast carcinoma. Of course, it is very uncommon.

The few patients I have seen with Jodbasedow disease, whether induced by SSKI, or more recently amiodarone, have had undetectable RAIU values. Nevertheless, it has been reported recently that some patients with amiodarone-induced thyrotoxicosis do not have suppressed RAIU values.[120] The difference might relate to iodide intake. The patients of Martino et al[120] who had normal RAIU values also had goiter and were from an area of "moderate iodine deficiency."

References

1. Griffin JE: Southwestern internal medicine conference: The dilemma of abnormal thyroid function test—Is thyroid disease present or not? Am J Med Sci 1985;289:78.
2. Borst GC, Eil C, Burman KD: Euthyroid hyperthyroxinemia. Ann Intern Med 1983;98:366.
3. Larsen PR: Thyroid-pituitary interaction. N Engl J Med 1982;306:23.
4. Braverman LE, Ingbar SH, Sterling K: Conversion of thyroxine (T_4) to triiodothyronine (T_3) in athyreotic human subjects. J Clin Invest 1970;49:855.
5. Hollander CS, Odak VV, Prout TE, Asper SP Jr: An evaluation of the role of prealbumin in the binding of thyroxine. J Clin Endocrinol Metab 1962;22:617.

6. Mendel CM, Cavalieri RR: Thyroxine distribution and metabolism in familial dysalbuminemic hyperthyroxinemia. J Clin Endocrinol Metab 1984;59:499.
7. Anderson BG: Free thyroxine in serum in relation to thyroid dysfunction. JAMA 1968;203:135.
8. Utiger RD: Decreased extrathyroidal triiodothyronine production in nonthyroidal illness: Benefit or harm? Am J Med 1980;69:807.
9. Howorth PJN, MacLagan NF: Clinical application of serum-total-thyroxine estimation, resin uptake and free-thyroxine index. Lancet 1969;1:224.
10. Sarin RK, Anderson BG: Serum thyroxine resin uptake of liothyronine I-125, and free thyroxine index. Arch Intern Med 1970;126:631.
11. Moses AC, Lawlor J, Haddow J, Jackson IMD: Familial euthyroid hyperthyroxinemia resulting from increased thyroxine binding to thyroxine-binding prealbumin. N Engl J Med 1982;306:966.
12. Stockigt JR, Topliss DJ, Barlow JW, et al: Familial euthyroid thyroxine excess: An appropriate response to abnormal thyroxine binding associated with albumin. J Clin Endocrinol Metab 1931;53:353.
13. Konishi J, Lida Y, Kousaka T, et al: Effect of antithyroxine autoantibodies on radioimmunoassay of free thyroxine in serum. Clin Chem 1982;28:1389.
14. Sakata S, Nakamura S, Miura K: Autoantibodies against thyroid hormones on iodothyronine. Ann Intern Med 1985;103:579.
15. Rajatanavin R, Fournier L, DeCosimo D, et al: Elevated serum free thyroxine by thyroxine analog radioimmunoassays in euthyroid patients with familial dysalbuminemic hyperthyroxinemia. Ann Intern Med 1982;97:865.
16. Sakata S, Komaki T, Nakamura S, Miura K: Measurement of free triiodothyronine in serum in the presence of autoantibodies to it. Clin Chem 1985;31:1252.
17. Marshall JS, Levy RP, Steinberg AG: Human thyroxine-binding globulin deficiency. N Engl J Med 1966;274:1469.
18. Refetoff S, Robin NI, Alper CA: Study of four new kindreds with inherited thyroxine-binding globulin abnormalities. J Clin Invest 1972;51:848.
19. Rosenberg IN, Ahn CS, Mitchell ML: Effects of anabolic steroids upon circulating thyroid hormone. J Clin Endocrinol Metab 1962;22:612.
20. Chopra IJ, Huang T-S, Hurs RE, et al: A competitive ligand binding assay for measurement of thyroid hormone-binding inhibitor in serum and tissues. J Clin Endocrinol Metab 1984;58:619.
21. Oppenheimer JH, Schwartz HL, Mariash CN, Kaiser FE: Evidence for a factor in the sera of patients with nonthyroidal disease which inhibits iodothyronine binding by solid matrices, serum proteins, and rat hepatocytes. J Clin Endocrinol Metab 1982;54:757.
22. Inada M, Sterling K: Thyroxine turnover and transport in Laennec's cirrhosis of the liver. J Clin Invest 1967;46:1275.
23. Fialkow PJ, Giblett ER, Musa B: Increased serum thyroxine-binding globulin capacity: Inheritance and linkage relationships. J Clin Endocrinol Metab 1970;30:66.
24. Staeheli V, Vallotton MB, Burger A: Detection of human anti-thyroxine and anti-triiodothyronine antibodies in different thyroid conditions. J Clin Endocrinol Metab 1975;41:669.

25. Rodriguez-Espinosa J, Ordonex-Llanos J, Gomez-Gerique JA, Mora J: Anti-triiodothyronine autoantibodies interfere with two recently commercialized radioimmunoassays for free triiodothyronine. Clin Chem 1984;30:151.
26. Alexander NM: Free thyroxin in serum: Labeled thyroxin-analog methods fall short of their mark. Clin Chem 1986;32:417.
27. Gow SM, Kellet HA, Toft AD, Beckett GJ: Accuracy and precision of five analog radioimmunoassays for free thyroxine compared. Clin Chem 1985;31:1868.
28. Csako G, Zwelg MH, Benson C, Ruddel M: On the albumin dependence of measurement of free thyroxin. I. Technical performance of seven methods. Clin Chem 1986;32:108.
29. Caldwell G, Gow SM, Sweeting VM, et al: A new strategy for thyroid function testing. Lancet 1985;1:1117.
30. Ross DS: New sensitive immunoradiometric assays for thyrotropin. Ann Intern Med 1986;104:718.
31. Mardell RJ, Gamlen TR, Winton MRJ: High sensitivity assay of thyroid stimulating hormone in patients receiving thyroxine for primary hypothyroidism and thyroid carcinoma. Br Med J 1985;290:355.
32. Wartofsky L, Burman KD: Alterations in thyroid function in sick patients with systemic illness: The "euthyroid sick syndrome". Endocrine Rev 1982;3:164.
33. Chopra IJ, Solomon DH, Hepner GW, Morgenstein AA: Misleadingly low free thyroxine index and usefulness of reverse triiodothyronine measurement in nonthyroidal illnesses. Ann Intern Med 1979;90:905.
34. Kaplan MM, Larsen PR, Crantz FR, et al: Prevalence of abnormal thyroid function test results in patients with acute medical illnesses. Am J Med 1982;72:9.
35. Chopra IJ, Hershman JM, Pardridge WM, Nicoloff JT: Thyroid function in nonthyroidal illness. Ann Intern Med 1983;98:946.
36. Kaptein EM, Robinson WJ, Grieb DA, Nicoloff JT: Peripheral serum thyroxine, triiodothyronine and reverse triiodothyronine kinetics in the low thyroxine state of acute nonthyroidal illness. J Clin Invest 1982;69:526.
37. Jansen M, Oostdijk W, Kingma BE, et al: Hyperthyroxinemia due to decreased peripheral triiodothyronine production. Lancet 1982;2:849–851.
38. Cooper DS, Daniels GH, Ladenson PW, Ridgway EC: Hyperthyroxinemia in patients treated with high-dose propranolol. Am J Med 1982;73:867.
39. Kleinmann RE, Vagenakis AG, Braverman LE: The effect of iopanoic acid on the regulation of thyrotropin secretion in euthyroid subjects. J Clin Endocrinol Metab 1980;51:399.
40. Burger A, Dinichert D, Nicod P, et al: Effect of amiodarone on serum triiodothyronine, reverse triiodothyronine, thyroxine and thyrotropin. J Clin Invest 1976;58:255.
41. Hamburger, JI, Hamburger SW: Diagnosis and management of large toxic multinodular goiter. J Nucl Med 1985;26:888.
42. Caplan RH, Glasser JE, David K, et al: Thyroid function tests in elderly hyperthyroid patients. J Am Geriatr Soc 1978;26:116.
43. Hamburger JI: Application of the radioiodine uptake to the clinical evaluation of thyroid disease. Semin Nucl Med 1971;1:287.
44. Bode HH, Danon M, Weintraub BD, et al: Partial target organ resistance to thyroid hormone. J Clin Invest 1973;52:776.

45. Refetoff S, DeWind LT, DeGroot LJ: Familial syndrome combining deaf-mutism, stippled epiphyses, goiter and abnormally high PBI: Possible target organ refractoriness to thyroid hormone. J Clin Endocrinol Metab 1967;27:279.
46. Refetoff S: Syndromes of thyroid hormone resistance. Am J Physiol 1982;243:E88.
47. Weintraub BD, Gershengorn MC, Kourides IA, Fein H: Inappropriate secretion of thyroid-stimulating hormone. Ann Intern Med 1981;95:339.
48. Spratt DI, Pont A, Miller MD, et al: Hyperthyroxinemia in patients with acute psychiatric disorders. Am J Med 1982;73:41.
49. Bouillon R, Naesens M, VanAssche FA, et al: Thyroid function in patient with hyperemesis gravidarum. Am J Obstet Gynecol 1982;143:922.
50. Goluboff LG, Sisson JC, Hamburger JI: Hyperthyroidism associated with pregnancy. Obstet Gynecol 1974;44:107.
51. Hamburger JI, Meier DA, Szpunar WE: Factitious evaluation of thyrotropin in euthyroid patients. N Engl J Med 1985;313:267.
52. Ingbar JC, Borges M, Iflah S, et al: Elevated serum thyroxine concentration in patients receiving "replacement" doses of levothyroxine. J Endocrinol Invest 1982;5:77.
53. Salmon D, Rendell M, Williams J, et al: Chemical hyperthyroidism. Arch Inter Med 1982;142:571.
54. Bantle JP, Oppenheimer JH, Schwartz HL, et al: TSH response to TRH in euthyroid, hypercholesterolemic patients treated with graded doses of dextrothyroxine. Metabolism 1981;30:63.
55. Morley JE, Shafer RB, Elson MK, et al: Amphetamine-induced hyperthyroxinemia. Ann Intern Med 1981;93:707.
56. Lundberg PA, Jagenburg R, Lindstedt G, Nystrom E: Heparin in vivo effect on free thyroxin. Clin Chem 1982;28:1241.
57. Frost PH, Mendel CM: Mechanism of the heparin-induced increase in free thyroxine in plasma. Abstract 22, American Thyroid Association 62nd Meeting, Washington, 1987.
58. Baker BA, Gharib H, Markowitz H: Correlation of thyroid antibodies and cytologic features in suspected autoimmune thyroid disease. Am J Med 1983;74:941.
59. Hamburger JI, Miller JM, Kini Sr: Lymphoma of the thyroid. Ann Intern Med 1983;99:685.
60. Brent GA, Hershman JM, Braunstein GD: Patients with severe nonthyroidal illness and serum thyrotropin concentrations in the hypothyroid range. Am J Med 1986;81:463.
61. Spencer C, Elgen A, Shen D, et al: Specificity of sensitive assays of thyrotropin (TSH) used to screen for thyroid disease in hospitalized patients. Clin Chem 1987;33/8:1391.
62. Sawin CT: Hypothyroidism. Med Clin N Am 1985;69:989.
63. Hamblin PS, Dyer SA, Mahr VS, et al: Relationship between thyrotropin and thyroxine changes during recovery from severe hypothyroxinemia of critical illness. J Clin Endocrinol Metab 1986;62:717.
64. Wartofsky L: Thyroiditis. Thyroid Clin 1981;1:1.
65. Hay ID: Thyroiditis: A clinical update. Mayo Clin Proc 1985;60:836.

66. Bastenie PA, Bonnyns M, Van Haelst L: Grades of subclinical hypothyroidism in asymptomatic autoimmune thyroiditis revealed by the thyrotropin-releasing hormone test. J Clin Endocrinol Metab 1980;51:163.
67. Cooper DS: Subclinical hypothyroidism. JAMA 1987;258:246.
68. Rosenthal MJ, Hunt WC, Garry PJ, Goodwin JS: Thyroid failure in the elderly. Microsomal antibodies as discriminant for therapy. JAMA 1987;258:209.
69. Maenpaa J, Raatikka M, Rasanen J, et al: Natural course of juvenile autoimmune thyroiditis. J Pediatr 1985;107:898.
70. Sklar CA, Qazi R, David R: Juvenile autoimmune thyroiditis. AJDC 1986;140:877.
71. Salmon D, Rendell M, Williams J, et al: Chemical hyperthyroidism. Arch Intern Med 1982;142:571.
72. Stock JM, Surks MI, Oppenheimer JH: Replacement dosage of L-thyroxine in hypothyroidism. N Engl J Med 1974;290:529.
73. Fish LH, Schwartz HL, Cavanaugh J, et al: Replacement dose, metabolism, and bioavailability of levothyroxine in the treatment of hypothyroidism. N Engl J Med 1987;316:764.
74. Hennessey JV, Evaul JE, Tseng YC, et al: L-thyroxine dosage: A reevaluation of therapy with contemporary preparations. Ann Intern Med 1986;105:11.
75. Pearce CJ, Himsworth RL: Total and free thyroid hormone concentrations in patients receiving maintenance replacement treatment with thyroxine. Br Med J 1984;288:693.
76. Utiger RD: Decreased extrathyroidal triiodothyronine production in nonthyroidal illness: Benefit or harm? Am J Med 1980;69:807.
77. Blank MS, Tucci JR: A case of thyroxine thyrotoxicosis. Arch Intern Med 1987;147:863.
78. Larsen PR, Silva JE, Kaplan MM: Relationships between circulating and intracellular thyroid hormones: Physiological and clinical implications. Endocr Rev 1981;2:87.
79. Spcnccr CA, Lai-Roscnfcld AO, Guttlcr RB, ct al: Thyrotropin sccrction in thyrotoxic and thyroxine-treated patients: Assessment by a sensitive immunoenzymometric assay. J Clin Endocrinol Metab 1986;63:349.
80. Martino E, Bambini G, Bartalena L, et al: Human serum thyrotrophin measurement by ultrasensitive immunoradiometric assay as a first-line test in the evaluation of thyroid function. Clin Endocrinol 1986;24:141.
81. Bayer MF, Kriss JP, McDougall IR: Clinical experience with sensitive thyrotropin measurements: Diagnostic and therapeutic implications. J Nucl Med 1985;26:1248.
82. Coindre J-M, David J-P, Riviere L, et al: Bone loss in hypothyroidism with hormone replacement. Arch Intern Med 1986;146:48.
83. Ahmann AJ, Solomon B, Duncan WI, Wartofsky L: Normal bone mineral density (BMD) in premenopausal women on suppressive doses of L-thyroxine. Abstract 41, American Thyroid Association 62nd Meeting, Washington, 1987.
84. Jennings PE, O'Malley BP, Griffin Ke, et al: Relevance of increased serum thyroxine concentrations associated with normal serum triiodothyronine values in hypothyroid patients receiving thyroxine: A case for "tissue thyrotoxicosis". Br Med J 1984;289:1645.

85. Franklyn JA, Sheppard MC: "Tissue thyrotoxicosis." Br Med J 1985;290:393.
86. Bell GM, Sawers SA, Forfar JC, et al: The effect of minor increments in plasma thyroxine on heart rate and urinary sodium excretion. Clin Endocrinol 1983;18:511.
87. Fraser WD, O'Reilly D St J, Thomson JA: Letter to the Editor. Br Med J 1985:290:1985.
88. Gow SM, Caldwell G, Toft AD, et al: Relationship between pituitary and other target organ responsiveness in hypothyroid patients receiving thyroxine replacement. J Clin Endocrinol Metab 1987;64:364.
89. Toft AD: Thyroxine replacement treatment: Clinical judgment or biochemical control? Br Med J 1985;291:233.
90. Ingbar JC, Borges M, Iflah S, et al: Evaluating serum thyroxine concentrations in patients receiving "replacement" doses of levothyroxine. J Endocrinol Invest 1982;5:77.
91. Volpe R: Reevaluation of L-thyroxine doses (Letter to Editor). Ann Intern Med 1986;105:798.
92. Wong MM, Volpe R: What is the best test for monitoring levothyroxine therapy? Can Med Assn J 1981;124:1181.
93. Bantle JP: Replacement therapy with levothyroxine: Evolving concepts. Thyroid Today 1987;10:1.
94. Ross DS, Neer RM, Ridgway EC, Daniels GO: Subclinical hyperthyroidism and reduced bone density as a possible result of prolonged suppression of the pituitary-thyroid axis with L-thyroxine. Am J Med 1987;82:1167.
95. Riesco G, Tavrog A, Larsen PR, Krulich L: Acute and chronic responses to iodine deficiency in rats. Endocrinology 1987;100:303.
96. Bigos ST, Ridgway EC, Kourides IA, Maloof F: Spectrum of pituitary alterations with mild and severe thyroid impairment. J Clin Endocrinol Metab 1978;46:317.
97. Selenkow HA, Ingbar SH: In: Wintrobe et al, eds. Harrison's Principles of Internal Medicine, 6th ed. New York: McGraw-Hill, 1970, p 456.
98. Evered D, Young ET, Ormston BJ, et al: Treatment of hypothyroidism: A reappraisal of thyroxine therapy. Br Med J 1973;3:131.
99. Snyder PJ, Utiger RD: Inhibition of thyrotropin response to thyrotropin-releasing hormone by small quantities of thyroid hormones. J Clin Invest 1972;51:2077.
100. Goolden AWG, Bateman D, Torr S: Red cell sodium in hyperthyroidism. Br Med J 1971;2:552.
101. Wilcox AH, Levin GE: Erythrocyte ouabain binding in patients receiving thyroxine. Clin Endocrinol 1987;27:205.
102. Bell GM, Sawers JSA, Toft AD, Doig A: Nocturnal natriuresis in hyperthyroidism. Clin Endocrinol 1982;16:177.
103. Sheridan P: Thyroid hormones and the liver. Clin Gastroenterol 1983;12:797.
104. Klion FM, Segal R, Schaffner F: The effect of altered thyroid function on the ultrastructure of the human liver. Am J Med 1973;50:317.
105. Ashkar FS, Miller R, Smoak WM, Gilson AJ: liver disease in hyperthyroidism. South Med J 1971;64:462.
106. Beckett GJ, Kellet HA, Gow SM, et al: Raised plasma glutathione S-transferase values in hyperthyroidism and in hypothyroid patients receiving thyroxine replacement: Evidence for hepatic damage. Br Med J 1985;291:427.

107. Paul T, Kerrigan J, Braverman LE, Baran D: Long-term L-thyroxine therapy is associated with decreased femoral neck bone density in premenopausal women. Clin Res 1987;35:648A (Abstract).
108. Editorial: Extrathyroidal actions of TRH. Lancet 1984®:560.
109. Borowski GD, Garofano CD, Rose LI, Levy RA: Blood pressure response to thyrotropin-releasing hormone in euthyroid subjects. J Clin Endocrinol Metab 1984;58:197.
110. Seth J, Kellett HA, Caldwell G, et al: A sensitive immunoradiometric assay for serum TSH—a replacement for the TRH test? Br Med J 1984;289:1334.
111. Kerr DJ, Alexander WD: Is the TRH test usually unnecessary? Lancet 1984®:1161.
112. Rendell M, Salmon D: "Chemical hyperthyroidism": The significance of elevated serum thyroxine levels in L-thyroxine treated individuals. Clin Endocrinol 1985;22:693.
113. Fraser WD, Biggart EM, O'Reilly DS, et al: Are biochemical tests of thyroid function of any value in monitoring patients receiving thyroxine replacement? Br Med J 1986;293:808.
114. Paul TL, Kerrigan J, Kelly AM, et al: Long-term L-thyroxine therapy is associated with decreased hip bone density in premenopausal women. JAMA 1988;259:3137.
115. Cooper DS: Thyroid hormone and the skeleton: A bone of contention. JAMA 1988;259:3175.
116. Melamed S, Geola FL, Reed AW, et al: A comparison of methods for assessing thyroid function in nonthyroidal illness. J Clin Endocrinol 1982;54:300.
117. Mendel CM, Frost PH, Kunitake ST, Cavalieri RR: Mechanism of the heparin-induced increase in the concentration of free thyroxine in plasma. J Clin Endocrinol Metab 1987;65:1259.
118. Edmonds CJ, Thompson BD: Hyperthyroidism induced by secondary carcinoma in the thyroid. Clin Endocrinol (Oxf) 1978;8:411.
119. Shimaoka K, Van Herle AJ, Dindogru A: Thyrotoxicosis secondary to involvement of the thyroid with malignant lymphoma. J Clin Endocrinol Metab 1976;43:64.
120. Martino E, Aghini-Lombardi F, Lippi F, et al: Twenty-four radioactive iodine uptake in 35 patients with amiodarone associated thyrotoxicosis. J Nucl Med 1985;26:1402.

5

Use of Modern Thyroid Tests in Illustrative Ambulatory Patients

JOEL I. HAMBURGER

The following patients were seen in my office in the early part of 1988 and they seemed to have laboratory findings that were not readily translated into therapeutic mandates. The first three cases address the issue of the relative usefulness of in vitro tests in the monitoring of LT_4 treatment of hypothyroidism. They are not isolated examples, but representative of many patients I see with similar data.

Unless otherwise stated the normal ranges for the various assays are:

Thyroxine (T_4): 4.5 to 11.5 μg/dl
Free thyroxine (FT_4): 0.6 to 2.2 ng/dl
Free triiodothyronine (FT_3): 2.4 to 4.9 pg/ml
Supersensitive thyroid-stimulating hormone (STSH): 0.3 to 4.0 μU/ml

Case 1. CC, a 24-year-old woman

10/87	Her gynecologist detected an elevated serum TSH level. He prescribed Synthroid, 0.2 mg daily. Initially she felt better, but after a few weeks she noted tremor, palpitation, and a 9-pound weight loss in the following 5 months.
2/5/88	She had fine warm skin, a resting heart rate of 120, hyperactive reflexes, a blood pressure of 130/60, and vigorous heart sounds. The FT_4 was 3.2; the T_4 was 12.5; the FT_3 was 4.6; and the STSH was $<$0.1.

Case 2. BM, a 39-year-old woman

1982	She was treated with ^{131}I for Graves disease. Subsequently she was maintained on Levothroid.
9/22/87	She was feeling well. The thyroid gland was impalpable. The FT_4 was 1.8; the T_4 was 10.6; the FT_3 was 3.6; but the STSH was $<$ 0.1. Her dose of Levothroid was reduced from 0.1 mg, 8 tablets per week to 0.1 mg, 7 tablets per week.
2/8/88	She felt no different on the 0.7 mg weekly dose of Levothroid. The FT_4 was 1.8; the T_4 was 13.4 (the patient was taking estrogen); the FT_3 was 3.2; and the STSH was 3.0.

For case 1 it seems clear that the dose of 0.2 mg daily of Synthroid was excessive. Would any of the panel members disagree? (All agreed.) However, those who rely on the T_3 level might say that no reduction in dosage is needed because the serum T_3 concentration is not elevated.

Comment by Dr. Leslie J. DeGroot

The FT_3 level although within your normal range is, nevertheless, at the upper level of normal, and one should not be a slave to the laboratory when the clinical picture is so clear.

Response by Dr. Joel I. Hamburger

Case 2 was free of symptoms of hyperthyroidism, and all tests were normal when she was rechecked in April of 1988 on 0.1 mg daily of LT_4. In September of 1987 all tests were normal except the subnormal STSH value. Furthermore, the patient felt well, and she felt no different after her dose of LT_4 was reduced. Would the panel agree that this dose reduction was proper?

Comment by Dr. John T. Dunn

It is hard to believe that such a small dose reduction of LT_4 would produce so much change in her STSH level.

Response by Dr. Ian D. Hay

I would have reduced her LT_4 dose.

Discussion by Dr. Joel I. Hamburger

Would anyone not have reduced the LT_4 dose in this patient? (No one disagreed with dose reduction.) Then let us now look at case 3, a further step in what will obviously be a progression to a less obvious situation.

Case 3. JG, a 31-year-old woman

3/4/77 This patient presented with a diffuse firm goiter twice normal size and symptoms of hypothyroidism. The T_4 value was 3.6 μg/dl (N = 5.4 to 11.5). The TSH value was 50 μU/ml. She has been treated with levothyroxine ever since.

2/12/88 She comes for a routine annual evaluation. She is feeling fine on 0.1 mg daily of Levothroid. The thyroid is normal in size. The heart rate is 80, and her weight is 133 pounds, down from 139 pounds the previous year (she was dieting). Her height is 65″. The FT_4 is 1.4; the T_4 is 7.8; but the STSH is 0.1 μU/ml.

What would you do here, rely on the implications of the subnormal STSH and reduce the LT_4 dose, or rely on the clinical findings and mid-normal values for FT_4 and T_4 levels?

Response by Dr. Ian D. Hay

I would reduce her LT_4 dose.

Comment by Dr. Leslie J. DeGroot

I would have her tested in another lab. I don't see this combination of test values in my patients.

Discussion by Dr. Joel I. Hamburger

Dr. DeGroot, this is not an isolated example, but she is representative of a sizeable number of patients on LT_4 with these values, this is why I present her case. When I encounter what seems to be an inappropriately subnormal STSH value in a young asymptomatic patient like this, with a FT_4 value that is not even close to the upper end of the normal range, it has been my practice to leave well enough alone. Of course, I am not sure that that is ideal treatment, and I hope that it may yet be possible with better TSH assays to differentiate patients with subnormal, but detectable, TSH levels from those with truly undetectable levels.

Case 4. RW, an 80-year-old woman

6/12/87 The patient was referred for evaluation of a painless midline neck swelling present for 40 years, but increasing in size in the past year. Her weight was stable at 111 pounds (59″ height), and other than occasional palpitation, muscle cramps, and paresthesias, there were no symptoms suggesting thyroid dysfunction. Her heart rate was 88 and the blood pressure was 160/80. The thyroid gland contained a 4 cm irregular midline mass, and a 1.5 cm nodule in the right lobe. A pertechnetate image of the thyroid showed very poor concentration of the tracer, most of which was within the midline mass. Fine-needle biopsy on both masses suggested nodular goiter. The FT_4 value was 0.7; the T_4 was 5.5; the FT_3 was 9.2; and the STSH was 1.1. Observation was advised.

2/2/88 Reevaluation revealed no change in symptoms or physical findings. Specifically, there had been no weight loss or other findings of hyperthyroidism. The FT_4 was 0.7; the T_4 was 6.5; but the FT_3 was 13.6; and the STSH was 1.2.

1. Is this developing T_3 toxicosis?
2. Is this preferential T_3 hypersecretion in compensation for reduced capacity for T_4 synthesis?
3. Would we be better off not to have obtained the FT_3 level?

Response by Dr. Leslie J. DeGroot

I have no experience with the FT_3 assay; I measure total T_3 when I want to know about T_3. Abnormal protein binding of T_3 is less common than with T_4.

Comment by Dr. Joel I. Hamburger

Nevertheless, if one is going to measure T_3 and if one can do a FT_3 assay, why not do the FT_3 assay? The cost is the same in my lab.

Response by Dr. Leslie J. DeGroot

If they cost the same, then I would agree. Could this patient have an elevated FT_3 because of T_3-binding antibodies, a common problem in Hashimoto's thyroiditis?

Comment by Dr. Joel I. Hamburger

We routinely add polyethylene glycol to our FT_3 assay system to eliminate interference with T_3-binding antibodies.

Response by Dr. James C. Sisson

It would appear that this is an unusual instance of impaired T_4 synthesis and preferential compensatory T_3 synthesis.

Response by Dr. John T. Dunn

I seem to recall that euthyroid goitrous patients with elevated T_3 levels are endemic in some areas of New Guinea.

Discussion by Dr. Joel I. Hamburger

This is another instance in which serum T_3 levels seemingly provide no inference of intracellular thyroid hormone activity. In sum, those situations are so common that it is difficult for me to place any reliance on T_3 levels in the monitoring of patients treated with LT_4.

Case 5. TT, a 24-year-old woman.

2/4/88 This patient was referred for hyperthyroidism of recent onset. History and physical findings were typical, including a diffuse goiter 3½ times normal size. The FT_4 was 3.6; the FT_3 was 13.8; and the STSH was 0.1. However, the radioactive iodine uptake (RAIU) was only 11%. On closer questioning she revealed that she had been taking a cough syrup that contained 15 mg of organically bound iodine per teaspoon, 4 teaspoons daily for 3 weeks.

What would you do in addition to stopping the cough syrup, assuming ^{131}I therapy were preferred?

1. Treat with antithyroid drugs for 4 weeks, and then give ^{131}I.
2. Treat with a beta blocker for 4 to 6 weeks, then the RAIU should be high enough for ^{133}I therapy.
3. Just give a higher dose of ^{131}I.
4. Recheck the RAIU in 1 week, because it will be high enough for a conventional dose of ^{131}I.

Response by Dr. Leonard Wartofsky

In about 1 week she will have a high RAIU.

Discussion by Dr. Joel I. Hamburger

In 1 week her RAIU was 57%. This case is exemplary of the principle that an iodine exposure does not block the RAIU in hyperthyroid patients with Graves disease for very long, and conversely, if the RAIU is blocked for very long the patient is not likely to be very hyperthyroid. Nodular goiter with Jodbasedow disease is an exception to this rule.

Case 6. SS, a 30-year-old woman

9/20/79 This patient was told that she had a goiter and hypothyroidism 6 months earlier by another physician. She had taken Synthroid 0.1 mg daily. The thyroid was normal in size and thyroid function tests were normal. Treatment was discontinued to determine whether it was still necessary.

1/24/80 The patient felt well. The T_4RIA was 7.3 μg/dl (N = 5.5 to 11.5); the free thyroxine index (FTI) was 2.0 (N = 1.4 to 4.0); and the TSH was 3.6 μg/ml. In response to thyrotropin-releasing hormones (TRH), the TSH level increased to 64 μU/ml. She was advised to take Synthroid 0.05 mg daily.

12/14/87 She had discontinued Synthroid 2 to 3 years earlier on her own. She now complained of fatigue and muscle cramps. There were no abnormalities on examination. The T_4 was 6.4 μg/dl (N = 4.5 to 11.5); the FT_4 was 0.7 ng/dl (N = 0.6 to 2.2); and the STSH was 3.6 μU/ml (N = 0.3 to 4.0). The microsomal antibody titer was positive at 1:1600. The STSH level 20 minutes after 100 μg of TRH was > 60 μU/ml. She was treated with 0.05 mg of Levothroid daily.

3/14/88 She was feeling much better. The T_4 was 10.0 μg/dl; the FT_4 was 1.1 ng/dl; and the STSH was 2.4 μU/ml.

I selected this case for discussion because Dr. Hay had said that he didn't find increased responses to TRH very helpful in identifying patients with hypothyroidism. I would like his comments on the following questions:

1. Do you agree this patient requires treatment with thyroid hormone?
2. How would you have known it without the TRH test?

Response by Ian D. Hay

Personally, I do not think that I would have treated a patient who "felt well" with a T_4 of 7.3 μg/dl and a TSH of 3.6 μU/ml. By contrast, in December 1987, we had a patient with symptoms whose T_4 was lower and who had a borderline FT_4 and a positive antibody titer. In this circumstance, I would object to a therapeutic trial aimed at improving the patient's symptoms.

As stated above, I may have treated the described patient on a trial basis in an attempt to improve her symptoms. I'm not sure whether a drop in this patient's TSH from 3.5 to 2.4 μU/ml would explain her renewed sense of well-being. I know that Tunbridge et al suggest that the presence of an elevated TSH and a positive MAb may result in hypothyroidism at a rate of only 3% to 5% per annum. I personally don't know what is the comparable risk for patients with positive MAb and only an exaggerated TSH response to TRH. There is a recently published report that suggests that such an exaggerated TRH response does not reliably predict future risk of hypothyroidism.

Discussion by Dr. Joel I. Hamburger

It seems that we will have to agree to disagree on this case. While in medical school I was impressed by the opinion of Nobel Laureate Dr. Jerome Conn who said that the burden of proof lies with the physician who allows the abnormal physiology to persist without correction when correction is simply accomplished. He had reference to hyperglycemia, but I think his comments are equally applicable to this case. I agree, however, that a final answer is not available.

Case 7. A 50-year-old woman previously treated for thyroid cancer

6/5/87 The patient had a total thyroidectomy for papillary carcinoma of the thyroid. She was subsequently treated with LT_4 0.1 mg daily.

12/14/87 She was feeling fine. Her FT_4 value was 1.6 ng/dl and her STSH value was <0.1 μU/ml.

4/4/88 The patient complained bitterly of fatigue, cold intolerance, and muscle cramps. Her FT_4 had dropped to 0.4 mg/dl and her TSH had increased to 57 μU/ml. She assured me that she had taken her thyroxine continuously and faithfully. Her only change was the addition of cholestyramine by her internist for treatment of hypercholesterolemia. After discontinuing the cholestyramine, her thyroid function tests returned to their levels of December 1987.

Since this patient was seen, two additional patients have been identified with the same problem, and for a third the problem was prevented when his wife called to ask if it was safe to take that drug with his thyroxine. The renewed and increasing concern for correction of hypercholesterolemia will make it likely that other patients dependent upon oral thyroxine will experience cholestyramine-induced hypothyroidism. Physicians need to be reminded that that drug interferes with the absorption of thyroxine from the gastrointestinal tract.

Section Summary: Diagnosis of Thyroid Function

JOEL I. HAMBURGER

In a book on diagnostic methods in clinical thyroidology, it is not surprising that the discussion of diagnosis of thyroid function would constitute the largest portion of the text. More surprising are the sharp differences of opinion between prominent authorities in this era of advanced assay technology. Some of these differences can be attributed, believe it or not, to lack of experience with the most advanced methods. In part, this lack of experience is the result of the lack of a perceived need for better assays because of relative comfort with the less precise but more familiar old test (e.g., the free thyroxine index). This brings us to the first issue for this summary.

What Are the Most Reliable Practical Assays for In Vitro Thyroid Function Tests?

In this discussion, it is presumed that the physician wants the information that the test provides. Whether this information might be needed depends upon the objective of the testing.

To measure the serum concentration of the fraction of T_4 that correlates best with thyroid hormone activity at the tissue level, the Clinical Assays Gammacoat two-step free T_4 assay is the best test. It provides appropriate information even in the presence of abnormal proteins that specifically bind T_4, as in the FDH syndrome; and it is the only assay that will do it. The free thyroxine index and other free T_4 assays that are analog based give falsely elevated values in the familial dysalbuminemic hyperthyroxinemia (FDH) syndrome, and so does the new immunochemiluminescent free T_4 (I know this because I tested the method on serum from a family with FDH). The Clinical Assays method is also the most reliable method in patients with nonthyroidal illness, although no free T_4 assay is completely reliable in these patients.

There is resistance to the use of this free T_4 assay for reasons in addition to a reluctance by some to part with an old friend (the FTI) even if that friend lets one down increasingly often. The principal additional objections are increased costs of the kits and increased technologists' time in performing the assays. In the face of intense pressure to provide high-volume laboratory procedures at the lowest possible costs, commercial laboratories are understandably reluctant to use a more costly test. Third-party payers do not provide extra compensation for high quality. One additional objection to the use of the free T_4 assay instead of the free thyroxine index (in which the value for the total T_4 concentration is available) is that one loses the ability to appreciate that there is abnormal T_4 binding (increased or decreased). If it is considered important to know this, one can simply perform a serum T_4 assay in addition to the free T_4

assay. The best test of serum protein T_4 binding capacity is the serum T_4 assay.

To measure the serum concentration of that fraction of T_3 that correlates best with thyroid hormone activity at the tissue level, a free T_3 assay is superior to an assay that measures the total T_3 concentration. There are two simple and reliable assays available that I have tested, the Amersham kit and the Diagnostic Products kit. Currently, I use the Diagnostic Products kit because it is simpler and less costly. It was interesting to me to observe that none of the other participants had experience with free T_3 assays. Since free T_3 assays can be done at no greater cost than total T_3 assays, I can think of no reason not to use the free T_3 assay routinely when information on serum T_3 concentration is desired.

If one wants to know the serum concentration of TSH, the best assay is clearly the new immunochemiluminescent assay (ICMA). This test gives a much sharper differentiation between subnormal and undetectable values than the latest supersensitive immunoradiometric assay (IRMA) methods. That ability to provide accurate data for very low concentrations of TSH is of critical importance if the need for TRH testing is to be eliminated in the diagnosis of hyperthyroidism. The introduction of the assay will meet resistance because of the need for new instrumentation and higher costs for the reagents. Both of these costs will fall as utilization of the assay increases.

What Is the Best Test for Screening Patients for Abnormal Thyroid Function?

There can no longer be any doubt that the best test is a supersensitive TSH assay, and the best assay is the ICMA assay. This assay gives an undetectable TSH level in hyperthyroidism; a normal, subnormal (but not undetectable), or slightly elevated level in euthyroid sick patients; and an elevated level in patients with primary hypothyroidism. Screening of ambulatory patients with clinical features suggestive of thyroid dysfunction with this single test would by highly cost effective. One may find subnormal values without hyperthyroidism in patients taking thyroid hormone. Undetectable values in patients taking thyroid hormone can safely be assumed to mean thyroid hormone overdosage, unless there is a reason to suppress TSH virtually completely (e.g., in the treatment of thyroid cancer).

It must be remembered that the principal purpose of a screening test is to exclude abnormal thyroid function. Normal and even subnormal ICMA TSH values seem likely to do this with great reliability. Undetectable values nearly always mean thyroid hormone excess, but one would then require additional clinical and laboratory data to differentiate the conventional hyperthyroid syndromes (Graves disease, toxic multinodular goiter, toxic autonomous nodules) from the spontaneously resolving hyperthyroidism of the subacute thyroiditis syndromes, Jodbasedow disease, iatrogenic or factitious thyrotoxicosis, and other less common syndromes associated with thyroid hormone excess.

Although an elevated TSH level raises the question of primary hypothyroidism, there are exceptions. Serum TSH levels may be slightly elevated in the elderly, in nonthyroidal illness, in early stages of Hashimoto's thyroiditis, or in other syndromes that can eventually lead to frank hypothyroidism. Before assuming that the patient is hypothyroid, and committing him or her to lifelong treatment with thyroid hormone, additional confirmatory clinical and laboratory data are needed. Of particular importance is a search for some etiology for the presumed hypothyroidism. In this regard, it is useful to remember that more than 95% of all primary hypothyroidism in the United States is the result of Hashimoto's thyroiditis, prior thyroidectomy, or prior radioactive iodine therapy. In the absence of these causes, one should make a diagnosis of hypothyroidism reluctantly and only after careful documentation. Whether an exaggerated response on TRH testing has a role in the confirmation of mild hypothyroidism or not may be argued. I find the test helpful in patients with evidence for Hashimoto's thyroiditis and borderline or slightly elevated TSH values, and for those with similarly borderline values, who are on thyroxine for bona fide primary hypothyroidism but still have persistent complaints that suggest (to them or to me) a need for a higher dose of thyroxine.

What Is the Best Test for Monitoring Patients Taking Thyroxine for Primary Hypothyroidism?

It surprised me to find that this issue proved so controversial. When physicians of the stature of Volpé, Oppenheimer, and others are inclined to place reliance upon the serum T_3 concentration, especially to avoid overdosage, and Larsen, Toft, and apparently the participants in this project rely on the supersensitive TSH assay, it might seem presumptuous of me to decide who is right. Nevertheless, sidestepping the arguments of whether an undetectable STSH value means thyroid hormone excess, and if it does whether that excess is harmful, there is another argument that militates against the use of a T_3 assay to monitor these patients. Clearly a normal T_3 does not exclude hypothyroidism. Hence a T_3 assay alone would be inadequate. At least a STSH assay would be needed. If the STSH assay alone were an adequate monitor, then the T_3 assay would be redundant and not cost effective. It is hard to argue that a STSH value well within the normal range is evidence that thyroid hormone therapy is appropriate. Similarly an elevated STSH value would nearly always mean that a higher dose of thyroid hormone was needed.

A subnormal but detectable STSH value might call for further evaluation by either free T_4 or free T_3 assay depending upon one's attitude. In my experience the free T_4 value is nearly always disproportionately elevated compared to the free T_3 value (when the thyroid hormone is thyroxine). Since I would reduce the thyroxine dose when the free T_4 is elevated, I do not need the free T;3 assay.

If the STSH value is undetectable, I would reduce the thyroxine dose regardless of the free T_4 or free T_3 levels. The only exceptions, of course, would be cancer patients or those with goiters or nodules for which full suppression of TSH might be the objective of the treatment.

The foregoing is the most cost-effective approach to monitoring thyroxine therapy. It is also consistent with the evidence, no matter how unconvincing to some authorities, that low-level overdosage of thyroxine may be harmful. Finally, it is consistent with the practical reality that there is no reason to think it necessary to give enough thyroxine to produce an undetectable STSH value in the treatment of hypothyroidism.

What Are the Best Tests to Rule Out Thyroid Dysfunction in Patients with Nonthyroidal Illness?

The heat generated by the interchanges on this subject was intense. Let us deal with the easier problem, that of excluding hyperthyroidism. There seems to be good evidence that the STSH value is, although often subnormal, not undetectable (especially by the ICMA assay) in sick patients without hyperthyroidism. If an undetectable STSH is found, the presumption of hyperthyroidism would be high. If free T_4, free T_3, and radioactive iodine uptake values are elevated (as will be the case most of the time in hyperthyroidism), the diagnosis will be easy. If the free T_4 value is normal and the free T_3 value is low (as would be anticipated in the euthyroid sick patient), hyperthyroidism would be too unlikely to be a significant contributing factor to the patient's debility to warrant treatment. The occasional problem patient might be one who happens to have the combination of a reproducibly undetectable STSH value, an elevated free T_4 value (possible in euthyroid sick patients as well as in hyperthyroids), a normal free T_3 (rather than the expected subnormal value), and a suppressed radioactive iodine uptake value because of recent iodide ingestion in radiographic contrast media or medications. This combination of findings should not be so common as to constitute a major diagnostic dilemma. After all, the patient might at least be cooperative enough to have a goiter, bilateral upper eyelid retraction, and hyperactive reflexes if he or she is truly hyperthyroid If those findings were all lacking, and the laboratory abnormalities were the only indicators suggesting hyperthyroidism, one might choose to observe, or start high-dose antithyroid drug therapy depending upon how critical the situation might be. I just do not believe this picture will appear very often.

The opposite problem is the critically ill patient with a T_4 of less than 5 μg/dl, and a slight elevation in TSH. Should thyroid hormone be given, as DeGroot maintains, or should we adhere strictly to the principle of "*primum non nocere*" as Wartofsky admonishes. Let us make the problem a bit stickier by adding that there is no goiter, antithyroid antibodies are negative, thyroid imaging shows a normal-appearing thyroid gland, there is no surgical scar on the neck, and it is known that the patient never received radioactive iodine therapy. This combination of circumstances focuses our attention on the issue of whether critical illness per se can impair the output or peripheral utilization of thyroid hormone. There is no final answer to this question. There is clear evidence of a high death rate in critical illness when the T_4 value is less than 5 μg/dl. It seems too improbable that a small dose of T_3, say 25 μg daily, would be harmful. It seems almost equally improbable that it would play a critical role in the survival of the patient. Wartofsky's question of why give T_4 if T_4 to T_3 conversion might be impaired seems logical to me. Of course, DeGroot gave T_4 because there is an intravenous preparation that might work more rapidly than oral T_3. Again, it is doubtful that that would be the critical element in the patient's survival. Secure in the knowledge that I will never have to face such a problem because I deal exclusively with ambulatory patients, I feel free to cast my vote for the 25 μg daily dose of T_3, believing that measures of unproved but possible efficacy are reasonable in desperate situations as long as they carry no reasonable risk of making matters worse.

By thus agreeing (at least partly) with both DeGroot and Wartofsky, I have followed the tradition of the great rabbinic sage of Chelm,* who when confronted with two litigants, after hearing the argument of the first promptly announced that he was right. The opposing litigant thereupon demanded to have his side of the argument heard, and after presenting his case the great sage promptly announced that he was right. The rabbi's wife, witnessing these proceedings, interrupted her husband to say that they both could not be right. To this irrefutable logic the great sage responded that she, too, was right!

---

*For the occasional reader who, even in this enlightened age, might not be fully conversant on the subject of Chelm, let me say that Chelm is a mythical village in eastern Europe whose inhabitants were widely known for being extraordinarily infatuated with intellectual absurdity.

Section II
Noninvasive Evaluation of Structural Abnormalities of the Thyroid

6

Clinical Usefulness of Serum Thyroglobulin Assays

JOHN T. DUNN

Thyroglobulin (Tg) is one of the major proteins of the thyroid. Its intimate association with the synthesis and storage of thyroid hormones has prompted extensive research over many decades. More recent work has shown that the serum also normally contains Tg, and that serum levels increase in a number of thyroid conditions, particularly differentiated cancer. These findings have led to numerous clinical studies, with occasionally conflicting results.

This article first reviews pertinent information about Tg, especially its presence in the serum, the factors influencing serum levels, and the nature of the assays used for its detection. It then focuses on studies of the serum Tg assay in differentiated thyroid cancer, and based on these reports, offers suggestions for its practical clinical application.

Thyroglobulin Structure and Function

Thyroglobulin has a molecular weight of 660,000 and a carbohydrate content of about 10%. Recent work has greatly advanced knowledge of its chemical structure (see references 1,2 for details and reviews). Briefly, the Tg gene codes for a polypeptide chain of ~2,750 amino acids synthesized on the rough endoplasmic reticulum of the thyroid cell. Post-translational events include glycosylation in the Golgi, association of two chains to form the 19S dimer of ~660,000 daltons, and iodination and hormone formation at the apical border of the cell. Iodide previously concentrated in the thyroid by active transport from the circulation is oxidized to iodine or some other transient intermediate through the action of a thyroidal peroxidase, and attached to tyrosyl residues that are already part of the structure of Tg. These initial iodinated products, monoiodotryosine (MIT) and diiodotyrosine (DIT) then couple within the molecular framework of Tg to form the thyroid hormones thyroxine (T_4) and triiodothyronine (T_3); two DITs couple to form T_4, and one DIT and one MIT to form T_3. The mature hormone-containing Tg is then stored in the follicular lumen as colloid. When thyroid hormone is required, lysosomal proteases digest Tg, the released thyroid hormones enter the circulation, and the nonthyroidal iodine and perhaps other constituents are reutilized within the thyroid cell.

Flexibility is an important feature of the Tg molecule. Most structural variation probably occurs posttranslationally, although we found changes in the amino acid composition of Tgs from experimental animals treated with thyroid-stimulating hormone (TSH).[3] The ability to alter the priority of iodination among Tg's hormonogenic sites is one example of its flexibility. We and others have described at least four major hormonogenic tyrosyls within the polypeptide chains of thyroglobulin.[4–6] The major T_4-forming site (site A) is close to the N-terminus of the molecule, and the major T_3-forming one (site C) is near the C-terminus. Another major hormonogenic site (B) is about 200 residues from the C-terminus, and appears to be particularly favored during early iodination or under conditions of low iodine availability.[7] Another site (D), in the molecule's midportion, assumes greater priority of utilization under TSH stimulation.[8]

Thyroglobulin structure also varies in certain thyroid diseases. We have reported significant differences in the amino acid composition and physical chemical characteristics of Tgs from nodular goiters[9,10] and follicular thyroid cancer.[11] Other cases in the literature have had low-molecular-weight Tg, difficulties in hormone synthesis despite adequate iodine availability, and decreased or absent amounts of Tg.[12,13]

Tg's antigenic structure has been of particular interest for its relation to thyroid autoantibodies in autoimmune thyroid diseases, its use in probing Tg's molecular structure, and its application to antibody production for immunoassays. Antibodies from autoimmune thyroid disease and from immunized rabbits can react to at least several T_4-containing sites.[14] Kohno et al[15] found that different monoclonals recognize different determinants on the Tg molecule, and that some clones show different reactivities to Tgs from different diseases, particularly papillary carcinoma. They concluded that different thyroid diseases might produce conformational changes in the Tg molecule. Similarly, Kurata et al[16] described six hybridomas that recognized different Tg epitopes. One group of these antibodies recognized structural features of the molecule unrelated to iodination; another group reacted with "iodination related epitopes other than iodoamino acids," and a final group appeared specific for T_4-containing portions of the molecule. In an immunohistochemical study, DeMicco et al[17] found that certain monoclonal antibodies showed greater reactivity with poorly differentiated anaplastic thyroid carcinomas than did their polyclonal antibody.

We have raised polyclonal antibodies to individual Tgs from normals, familial goiters, and Graves disease.[18] When tested against individual Tg samples, the three antibody preparations behaved similarly, suggesting they were recognizing similar epitopes. However, different Tgs, when used against any of these three antisera, showed wide variation in their degree of immunoreactivity. The following factors were associated with a decrease in immunoreactivity: the absence of proteolytic inhibitors in the Tg isolation procedure, the 27S species of Tg, and some Tgs from thyroid neoplasia and autoimmune thyroid disease. Of particular interest, the antiserum prepared with Tg from Graves disease contained antibodies not found in antisera against normal Tg or against Tg from a familial goiter, and this antiserum had a higher immunoreactivity with Tgs from Graves disease than with those from other conditions. This finding suggested that the antigenic sites of Tgs from Graves disease differed from those of normals or other thyroid diseases.

Thyroglobulin in Serum

Until about 20 years ago it was thought that Tg was confined to the thyroid gland. The presence of antibodies to Tg in autoimmune thyroid diseases was generally attributed to thyroid membrane defects, which would allow Tg leakage into the serum and provoke an antibody response. The development of radioimmunoassays as sensitive tools for detecting minute amounts of proteins and other substances led to the observation that Tg is a normal constituent of human and other mammalian serum.[19] A number of laboratories, particularly those of Van Herle,[19] Schneider,[20] and Pinchera,[21] have investigated the characteristics of serum thyroglobulin. The density of circulating thyroglobulin is somewhat less than that of thyroidal thyroglobulin, and from this Schneider[20] has suggested that circulating thyroglobulin is very poor in iodine content. He could not find differences in its carbohydrate composition or its peptide portion, and concluded that the main difference between serum and thyroidal Tg is in iodine content.

Circulating Tg clearly originates from thyroidal Tg.[22] Occasionally Tg may simply leak through defective or disrupted membranes; in such conditions, the serum Tg should be fairly representative of follicular Tg, and thus have a similar iodine content. Ikekubo et al[23] have postulated that some newly synthesized Tg may not reach the apical membrane, where iodination occurs, but instead is released directly into the circulation. Herzog[22] has studied Tg transport in an experimental model with inverted follicles, and concluded that Tg can be released by transcytosis, which would carry it across the thyroidal cell in vesicles formed from the apical membrane at the bases of microvilli, and thus the Tg would escape lysosomal degradation.

TSH stimulates the release of Tg into the circulation. Administration of one unit of bovine TSH produced a mean increment of 171% above baseline in seven normal subjects, with the peak occurring about 24 hours following injection.[24] However, endogenous TSH release by thyrotropin-releasing hormone (TRH) administration leads to a much earlier increase in serum Tg than is found after exogenous bovine TSH;[25] the reasons for this discrepancy are not clear. Bovine TSH also increased circulating Tg in rats.[26] In those studies the unstimulated serum Tg had a low iodine level, as deduced from its density on ultracentrifugation, but with TSH stimulation the density rose to a level approximating that of thyroidal Tg. This important observation suggests that the increase in Tg following TSH stimulation derives directly from the follicular Tg pool, rather than from enhancement of the poorly iodinated pathway normally responsible for serum Tg.

Direct release of Tg from diseased thyroids can also elevate serum Tg levels. Gebel et al[27] convincingly documented this process in a study of simple goiter, in which cell necrosis led to colloid leakage into interstitial spaces; these same patients had elevated serum Tg levels. Many other conditions associated with goiter, hyperplasia, or membrane disruption routinely elevate the serum Tg. Examples of these are Graves disease, endemic goiter, and subacute thyroiditis. Estimates of the half-life of serum Tg range from 1 to 4 days.[24]

Assays for Tg

Van Herle et al[28] described the first practical radioimmunoassay for Tg in humans. Most subsequent descriptions of serum Tg levels have used the same general technique. The antigen was Tg from humans undergoing thyroidectomy for Graves disease. It had been purified by two gel filtration steps, then injected into rabbits for induction of antibodies. The radioimmunoassay was carried out by double-antibody technique using ^{131}I-labeled hTg. The reported sensitivity was 1.6 ng/ml, and the mean concentration in normal subjects was 5.1 ng/ml.

Many other laboratories have developed their own radioimmunoassays for Tg. These were apparently prepared with whatever human thyroid tissue was convenient, usually surgical specimens of Graves disease or simple goiter. Careful attention to the details of different clinical studies shows some major differences in the sensitivity and, to some degree, the specificity of the antisera. A 1985 study reported the results of testing three standardized Tg samples in each of 37 laboratories in 18 countries.[29] The first standard, 5.3 ng/ml, was detectable in only 19 assays. The second, 30.6 ng/ml, was still not detected in three assays. The third standard, at 80.6 ng/ml, was detected by all. Both of the two higher samples gave mean values among the participating laboratories that were considerably less (by about 50% or more) than in the original standard. The most important possibility for such variation is probably the antisera produced among different laboratories. Other technical factors, such as the use of ^{125}I versus ^{131}I, lyophilization, and iodination technique, probably all come into play. Whatever the reason, the variation in assay sensitivity is a particularly sobering fact, and may be responsible for many of the conflicting conclusions from different studies. For the clinician, the most important advice is to send samples to a laboratory, usually a commercial one, with a well-established radioimmunoassay (RIA) or immunoradiometric assay (IRMA) program and with an assay that is sensitive to 1.0 ng/ml.

Surprisingly little attention has been paid to the antigen used for producing Tg antibodies for clinical assay. I have already described above the considerable variation in Tg molecules from different individuals and in different thyroid diseases, including thyroid cancer. These structural differences in turn lead to wide differences in the antigenic character of Tg. Recent studies with monoclonal antibodies have emphasized these differences.[15,16] At least some antibodies are directed at the iodinated sites found in normal glandular Tgs, although these sites may be absent in most circulating Tg. Despite these considerations, most papers dealing with preparation of antibodies give scant attention to the source of the Tg antigen. Since thyroidal Tg may not be the best antigen to reflect circulating Tg levels in thyroid cancer, much more attention should be given to the proper characterization of the antigens of greatest relevance to the clinical disease, and to the use of monoclonal antibodies specific for these antigens.

Clinical Use of Serum Thyroglobulin Measurement

By far the most important clinical application of serum Tg measurements is in patients with differentiated thyroid cancer. However, its use is not confined to thyroid cancer. In general, the serum Tg may be helpful in any circumstance where one wishes to demonstrate the presence, and particularly changes, in functioning thyroid tissue.[21,30] The following are some examples:

1. Hyperthyroidism. The serum Tg is usually markedly elevated in Graves disease and in subacute thyroiditis, but suppressed in thyrotoxicosis factitia. Usually these conditions can be distinguished by their typical clinical features, but occasionally not, and then the serum Tg can help.
2. Neonatal hypothyroidism. The clinical choice may be between thyroidal aplasia and a familial biochemical defect, without an obvious goiter. The serum Tg will typically be suppressed in the former but elevated in the latter.
3. Cysts. Most palpable cysts associated with the thyroid are in fact thyroidal, and this is evident from the rest of the clinical picture. Occasionally, palpable parathyroid cysts occur and need to be distinguished from thyroid cysts. Parathyroid cysts typically have a very clear watery

fluid quite different from thyroid cysts. This can be biochemically confirmed by assays for Tg and for parathormone. If the cyst is from the thyroid, the former will be high and the latter low; if from the parathyroid, the reverse.

The above are only several uses for the Tg radioimmunoassay. In general it can be used whenever a thyroid origin for a body fluid is questioned.

Thyroid Cancer

In 1975 Van Herle and Uller[31] clearly established that serum Tg is a useful marker for differentiated thyroid cancer. Their initial publication showed that untreated patients had elevations ranging from 22 to 445 ng/ml, that the serum Tg returned to normal in seven of these patients after thyroidectomy, that patients with medullary carcinoma did not have elevated values, that patients without residual thyroid or cancer had suppressed Tg levels, and that patients with documented metastases had marked elevations. Many publications have followed and amply confirmed all of these observations. The serum Tg has become a recognized and indispensable marker for following patients with differentiated thyroid cancer.

I will next review some of the more important clinical investigations of serum Tg and thyroid cancer. This list of studies makes no attempt to be all-inclusive, but does present a representative sample of available information. The following paragraphs describe each study briefly. Table 6.1 is a summary of some of the major features. There are two particularly important clinical questions regarding the use of serum Tg in thyroid cancer. Both refer to the patient who has had a total thyroidectomy for differentiated carcinoma, is now on T_4, and has a suppressed serum Tg (less than 1 ng/ml). The questions are: 1) does she need to have the serum Tg repeated while off T_4?, and 2) does she need a WBS (whole body scan)? Table 6.1 presents some conclusions (not necessarily those of the studies' authors) to these questions based on the information given in each paper.

1. Ashcraft and Van Herle.[32] Thirty-two patients had a total of 53 studies. All had had well-differentiated thyroid cancers removed by near-total or total thyroidectomy. They were grouped as follows: group 1, surgical ablation, no subsequent ^{131}I ablation; group 2, thyroidectomy, ^{131}I therapy, but subsequent positive or equivocal whole body scan; and group 3, thyroidectomy, ^{131}I ablation, and negative WBS. In all patients the exogenous T_4 therapy was withdrawn and the patients retested 4 weeks later. No patient with a Tg of less than 1 ng/ml on T_4, or less than 10 ng/ml off T_4, had detectable cancer. In the patients with total thyroid ablation, the serum Tg was more sensitive than the WBS in detecting

TABLE 6.1. Some major studies on serum Tg in thyroid cancer.

Authors and references:	Year	No. pts.	Assay sensitivity (ng/ml)	Conclusions to question	
				#1*	#2*
1. Ashcraft & Van Herle[32]	1981	32	<1	No	No
2. Black et al[33]	1981	274	1	No	No
3. Charles et al[34]	1980	48	<1	No	No
4. Echenique et al[35]	1982	68	15	NA	NA
5. Barsano et al[37]	1982	54	2–4	Yes	No
6. Feldt-Rasmussen et al[38]	1983	72	<5	NA	NA
7. Hufner et al[39]	1983	68	5	Yes	No
8. Ericsson et al[40]	1984	262	2	No	No
9. Pacini et al[41]	1985	82	3	Yes	NA
10. Valimaki and Lamberg[42]	1985	52	3	No	No
11. Girelli et al[43]	1986	291	3	Yes	No

*Questions refer to a patient with total thyroidectomy for differentiated cancer, now on T_4, and Tg is <1 ng/ml: 1) Does she need a repeat serum Tg off T_4?; 2) Does she need a whole body scan? NA = not applicable.

residual cancer; six patients had documented metastases with elevated Tg levels but negative WBS.

The authors also reviewed the published data on 1,323 cases of well-differentiated thyroid cancer in which the Tg and WBS were compared. Of 284 with metastases, 13% were missed by the WBS. Four percent (12 cases) had positive scans but undetectable TGs; of these, the assay may have been technically inadequate in nine. The authors recommended that a serum Tg value of less than 1 ng/ml while on T_4, or less than 10 ng/ml while off, removes the need for WBS unless there are other clinical factors to raise suspicion of metastases.

2. Black et al.[33] The group consisted of 274 patients, of whom 266 had previous thyroidectomy and 183 had had additional ablative radioiod ine. Of patients on T_4, the serum Tg was less than 5 ng/ml in three patients who subsequently had cancer, and was below this level in 128 without cancer. Twenty-six patients with cancer and one without had a serum level greater than 5. Of patients off T_4, six with cancer and 85 without cancer had Tg levels below 5, and 51 with cancer and 22 without had levels above 5. Antithyroglobulin antibodies did not affect the assay of serum Tg, in contrast to other studies. This point emphasizes that different antigens and different antibodies can be expected to produce different antisera, and this fact must be remembered in comparing different studies. This report did not specify the source of the Tg for antiserum production. The authors concluded that the serum Tg has its maximal diagnostic value while the patients are on suppressive T_4. They recommended reconsidering the practice of with-

drawing T_4 and scanning patients in follow-up, and would perform scans only when the Tg level is greater than 5.

3. Charles et al.[34] The study group consisted of 48 patients with differentiated thyroid cancer, after eight were excluded because of interfering antibodies. All had serum Tg assays while on T_4, and subsequent scans after T_4 withdrawal. Forty-eight other patients, without known thyroid disease, were used as controls for the assay; their mean normal Tg level was 16.5 ± 8.1 ng/ml, with a range of 3.4 to 35.1. In the study group, 24 patients had negative iodine scans and a mean Tg level of 3.6 (range < 0.06 to 6.7 ng/ml). In another 16 patients the scan was positive but confined to the thyroid bed; in this group the mean Tg level was 7.7 ± 4.8 ng/ml (range < 0.6 to 18.3). Eight patients had evidence of persistent thyroid cancer by scan; their mean Tg level was 33.2 (range 11 to 59). The authors recommended a cutoff of 7 ng/ml as discriminating. By this criterion, there were no false positives (increased Tg but negative scan) but 19% false negatives (positive scan with a low Tg); however, none of these "false negatives" showed evidence of carcinoma, and the positive scan presumably reflected residual thyroid tissue. The normals had higher Tg levels than in some other studies, a result attributed by the authors to differences in Tg purification, since the Tg itself came from the same laboratory as in the studies of Van Herle. They concluded that in their assay a Tg level of less than 7 ng/ml suggests no active carcinoma, whereas a higher level would be the basis for recommending a scan.
4. Echenique et al.[35] The study group contained 58 patients with differentiated thyroid cancer, all treated with prior surgery, some with radioiodine, and off thyroid suppression at the time of examination. Results of serum Tg determinations and whole body scans were compared. The sensitivity range of the Tg assay was 20 to 1,000 ng/ml, and the manufacturer reported the normal range as being up to 60 ng/ml. Another study has suggested that this particular assay kit was less satisfactory than a Stanford reference assay.[36]

   Thirteen patients had Tg levels greater than 60 ng/ml. All had evidence of metastatic disease even though three had normal ^{131}I scans. Of 16 patients with Tg levels between 20 and 60 ng/ml, two had lung metastases and two had regional lymph node metastases. Of 39 patients with Tg levels below 20 ng/ml, one had lung metastases and three had neck metastases. The authors recommended that the Tg level be used as a complement to the ^{131}I WBS but not as a substitute for it.

   This study suffers from having a relatively insensitive Tg assay. Even so it missed only four of 39 metastases, of which three were in the neck. The absolute serum Tg values were not given for these four patients. It is quite possible that use of a more sensitive assay would have improved the discrimination.

5. Barsano et al.[37] This retrospective study of differentiated thyroid cancer included 34 patients treated with thyroidectomy alone and 38 with thyroidectomy followed by radioiodine. The radioimmunoassay was a modification of the method of Van Herle, with a sensitivity of 2 to 4 ng/ml; normal levels were considered to range from 2 to 20 ng/ml. The source of the human Tg was not reported. In the group treated with surgery alone, 28 had positive uptake in the thyroid bed on WBS. Fifteen patients had Tg levels of less than 10 ng/ml and a positive scan, but in each case the uptake was found exclusively in the thyroid bed. Of 38 patients treated with both surgery and radioiodine, 22 had negative scans and Tg levels below 10 ng/ml. Of these, 18 showed undetectable Tg levels. Six patients with elevated Tgs had known metastatic disease, although two were not detectable by WBS. Nine patients had elevated Tg levels without evidence of cancer but five of these later showed either metastases or persistent neck uptake of ^{131}I. In a small subset of 20 patients, the serum Tg level was much lower when the patient was on replacement T_4. The authors concluded that a normal Tg level means there will be no uptake on WBS.
6. Feldt-Rasmussen et al.[38] The study group was 72 patients with differentiated thyroid cancer, all treated with thyroidectomy and ^{131}I and/or external irradiation. The Tg assay had a sensitivity slightly less than 5 ng/ml. Seventy-two control subjects without thyroid disease had a median Tg value of 17.9 ng/ml, with a 95% confidence limit of 6.8 to 44.0. While on T_4, 13 patients had a raised serum Tg. Of these, seven had known metastases, five later showed metastases, and a final patient could not be further evaluated. Of 43 patients with normal Tg levels, 41 had no metastases; the two remaining patients had metastases and elevated serum Tg levels when T_4 was withdrawn. The authors measured Tg antibodies, and showed that in some cases these may be associated with low or unmeasureable serum Tgs in patients with metastatic disease.
7. Hufner et al.[39] The study group consisted of 68 patients, all with documented metastases of differentiated thyroid cancer and treated with total thyroidectomy and subsequent ^{131}I. The Tg RIA had a sensitivity of 5 ng/ml. Of 31 patients with positive WBS, 26 had elevated serum Tg levels, and five were below 10 ng/ml. Of nine patients who were WBS negative, eight had elevated serum Tgs. Twenty-eight patients developed new metastases during follow-up. Of these, 24 were ^{131}I negative, and within that group, 18 had elevated Tgs. Four of the 28 patients were ^{131}I positive, and all showed elevated serum Tgs. The authors conclude that the serum Tg is the most useful measure in follow-up of patients with differentiated thyroid cancer after total thyroid ablation. Serum Tg levels were compared in 21 patients both on and off T_4 suppression; in most, the serum Tg increased when off T_4, but only two changed from a normal to an elevated Tg.

8. Ericsson et al.[40] This study included 262 patients with differentiated thyroid cancer. Most Tg measurements were while the patients were on T_4. The median serum Tg concentration in 400 healthy female controls was 13 ng/ml, with a range from nondetectable to 77. A median suppressed value for patients without normal or malignant thyroid tissue was 3 ng/ml, and the 97.5th percentile was 10 ng/ml, which was used as an upper limit for normal after total thyroidectomy. All 14 patients with metastatic cancer after thyroidal ablation had serum Tgs greater than 10 ng/ml, with the highest value in patients with bone metastases. Of 70 patients who had follicular carcinoma with no metastases, only nine had Tg levels greater than 10 ng/ml, and of these seven had residual thyroid tissue on WBS. All five patients with distant metastases of papillary carcinoma after initial treatment had elevated serum Tgs. One hundred twenty patients had papillary cancer without evidence of metastases after ablation. Of these, 104 had normal serum Tg levels. The remaining 16 all had residual uptake in the thyroid bed. Two patients had elevated serum Tg levels but negative scans. There were none with a positive scan and negative Tg. The only patients with metastases and normal serum Tg levels had medullary carcinoma. The authors concluded that the serum Tg level can replace the WBS for most patients who have had thyroidal ablation for differentiated thyroid cancer.
9. Pacini et al.[41] The patient group consisted of 82 patients with differentiated thyroid carcinoma, either with or without subsequent ablative treatment. The serum Tg was measured by an IRMA with a sensitivity of 3 ng/ml. Of 54 samples in 40 patients with persistent residual thyroid, 40% showed an elevated serum Tg while off T_4, and 14% while on. Of 44 samples from 11 patients with lymph node metastases, all had elevated serum Tg levels off T_4, while seven of 22 on T_4 showed suppressed Tg in the presence of lymph node metastases. Of patients with bone or lung metastases, none showed normal Tg levels when off T_4, and four of 58 showed low values while on it. The authors concluded that the determination of Tg off T_4 is a more accurate indicator of the presence of metastases than while on it.
10. Valimaki and Lamberg[42] This study included 52 patients with differentiated thyroid cancer, of whom 23 received ^{131}I in addition to radioiodine ablation, and two received external irradiation. The serum Tg was measured both on and off T_4 treatment, and the results compared with whole body scanning. The reference range for normal subjects in this Tg assay was less than 50 ng/ml. Of 24 patients with negative scans, 20 had undetectable Tgs while on T_4. Of these, the serum Tg became measurable in two subjects while off T_4. Two additional patients had normal serum Tg levels (39 and 15 ng/ml, respectively) that increased markedly on withdrawal of T_4. Twenty-eight patients had positive WBS, but only two of these had uptake outside

the thyroid area. Of these 28, 18 had undetectable Tgs on T_4, and when measured off T_4, the levels rose to the normal range (4 to 21 ng/ml), which was attributed to the remaining normal tissue. From their experience the authors note that in only one of the 38 patients with a suppressed Tg value while on T_4 was anything achieved by withdrawing T_4 and doing further studies. However, when patients had detectable Tg levels on T_4 therapy, further investigation off T_4 was recommended to detect possible metastases.

11. Girelli et al.[43] This report included 291 patients with differentiated cancer. All had had thyroidectomies and most had received ^{131}I. The Tg assay was done both by RIA and IRMA. One hundred sixty-six patients showed no evidence of metastases, and of these only three had detectable Tg levels. Of 66 patients with demonstrable metastases, the serum Tg was suppressed in 19 while on T_4, but the metastases were limited to neck nodes in 17 of these 19. When off T_4, all but two of these 66 patients had elevated Tg levels. Thus, the authors found that 8.4% of the patients had a false negative Tg level, but of these, 84% had only nodal metastases. The WBS for this group showed a much higher (31.8%) false negative rate. Since the incidence of false negatives was greatly reduced when the patient was off T_4, they recommended that serum Tg levels be carried out both on and off T_4.

These eleven studies help us answer the two questions posed above: 1) If a patient with a history of differentiated thyroid cancer has a serum Tg level of less than 1 ng/ml while on suppressive thyroid therapy, does he or she need to have further evaluation off T_4?; and 2) Does the same patient with the suppressed serum Tg need a whole body scan?

I believe the answer to both questions is "no." For the first question, the studies that used the most sensitive assays found virtually no detectable cancer when the serum Tg was suppressed, even while on T_4 therapy. Several of the studies that would advocate withdrawal of T_4 before relying on a serum Tg either used an insensitive assay or chose an arbitrary cutoff between normal and abnormal that may have been too high. The answer to this question is very important to the patient. The periodic hypothyroidism associated with withdrawal of thyroxine therapy is highly disruptive, and will involve a month or so of impaired function each time such testing is done. In addition, since these cancers are sensitive to TSH stimulation, it is probable that the increased TSH stimulation during thyroxine withdrawal may promote cancer growth.

These studies are even more supportive of a "no" answer to the second question, and show that the serum Tg level is a more sensitive indicator of residual thyroid cancer than is the whole body scan. It is obviously cheaper and simpler as well.

From these considerations, I conclude that a suppressed serum Tg level in a sensitive assay is good assurance against the continued presence of differentiated thyroid cancer. In the absence of other clinical features to suggest the presence of thyroid cancer, the best treatment is to keep the patient on T_4 and not withdraw it for further study. It should be emphasized that the serum Tg, like any laboratory test, needs to be integrated with the rest of the clinical picture, and other concerns in a particular patient may alter the recommendations just given.

Valuable as it is, the serum Tg tests could be improved. Better knowledge of the chemical and antigenic structure of Tg, and how it may differ in various thyroid diseases, could lead to selective monoclonal antibody production with the hope of greater specificity for serum Tg, particularly in thyroid cancer. More selective antibodies could also offer an opportunity to bypass the problems of patients who already have circulating Tg antibodies as part of autoimmune thyroid disease.

Summary

Tg is a normal component of human serum. It originates from the thyroid and probably reaches the circulation by transcytosis usually and by direct leakage occasionally. TSH stimulates its release from the thyroid. It can be detected by RIA or IRMA. Serum Tg determinations are helpful in many clinical situations, such as subacute thyroiditis, neonatal hypothyroidism, thyrotoxicosis factitia, and thyroid/parathyroid cysts. However, the most important clinical application by far is in the follow-up of differentiated thyroid cancer. Detailed analysis of several series leads to the following conclusions: 1) The serum Tg is an excellent marker for residual differentiated thyroid cancer after total thyroid ablation; 2) The assay should be sensitive to a level of 1 ng/ml for optimal use; 3) In a patient with a serum Tg of <1 ng/ml while on T_4, and no special clinical concern of cancer recurrence, it is not necessary or desirable to withdraw T_4 in order to repeat the serum Tg or to do whole body scanning; 4) The serum Tg is better and simpler than the whole body scan in following differentiated thyroid cancer.

Recent work on the chemical and immunologic structure of Tg offers exciting possibilities for developing antibodies that are highly specific for the serum Tg molecules found in thyroid cancer and other diseases. Such investigations could improve the applicability of the serum Tg assay in clinical thyroidology and should be pursued vigorously.

Commentary by Dr. Leslie J. DeGroot and Dr. Buyuni Jahazi: We Believe That Available Data Do Not Indicate That Serum Tg Assay Can Substitute for Whole Body ^{131}I Scans

The data reviewed by Dr. John Dunn are of great importance in the management of our patients with thyroid malignancy in the follow-up period after surgery and/or radioactive iodide administration. If serum Tg assay can be safely substituted for whole body ^{131}I scan, it is certainly crucial that this change be made in our management. However, if serum Tg assay has a degree of inaccuracy that is more than the physician or patient is willing to accept, then it would be an error to rely upon it. We have reviewed the papers that form the basis of Dr. Dunn's argument, and the conclusion we are able to draw is different from the position he has taken in his summary, and we must respectfully disagree with his conclusion.

The question of importance is whether, in patients with a negative physical exam and negative chest X ray, a serum Tg during replacement therapy, at the level believed to be appropriate for management of thyroid carcinoma, and which is within the accepted suppressed range, indicates the absence of malignancy. And second, whether an elevated serum Tg on suppressive therapy indicates the presence of malignancy. These are the important questions in management, since the answers dictate whether the patient should be made hypothyroid and given a radioactive iodide scan. There are other questions about the relationship of serum Tg and isotope scans that are of interest in the pathophysiology of cancer but not crucial in the management of patients. For example, it is clear that some patients have elevated serum Tg in the presence of metastases that do not take up radioactive iodide. However, this is not an argument against the use of the scans, since it is mandatory, in the presence of an elevated Tg and possible presence of malignancy, that a scan be performed to rule in or out the possibility of radioactive iodide therapy.

We have listed the studies and the data that bear on the two questions indicated above in Table 6.2. Most, but not all, of the studies indicate that patients on replacement therapy were given adequate doses of thyroid hormone to suppress TSH into the normal range, or in some instances, to prevent TRH responsiveness. Future studies of this problem should clearly indicate both the dose of thyroid hormone given, and the measurement taken as adequate suppression of TSH. Assays should be made by currently available supersensitive TSH methods. And, of course, there is the unsettled question as to whether the TSH value should be at the bottom end of the normal range, in the suppressed range, or zero, including suppression to TRH stimulation. Differences in degree of TSH stimulation could well account for variations in the level of Tg found in patients.

Most of the assays indicate sensitivity. Dr. Dunn indicates that the Tg should be measured by an assay sensitive to 1 ng/ml. Only one of the 13 studies reviewed suggests that their assay is this sensitive. Most indicate they are sensitive to between 2 and 5 ng/ml. Dr. Dunn states that a negative assay is one that is below 1 ng/ml. Only one of the assays uses this cutoff. The upper limit of an acceptable, suppressed TSH in these reports varies from 2.5 to 20 ng/ml. Most are in the range of 2.5 to 10 ng/ml.

Among patients who have presumed normal residual thyroid tissue and no known cancer, on suppressive therapy, 7% to 54% have Tgs above the author's suppressed value. Clearly, only a fraction of patients with

TABLE 6.2. Selected results of Tg assays in thyroid cancer patients.

Authors and reference	T_4Rx	TSH level (μU/ml)	Assay sensitivity (ng/ml)	Assay +	Patients with presumed normal residual thyroid and no known carcinoma		Patients with *no* residual thyroid and status unknown		Patients with definite carcinoma		Comment
					On T_4	Off T_4	On T_4	Off T_4	On T_4	Off T_4	
Schlumberger et al[44]	0.1 mg/m²	<5	2.5	>2.5	19+/35 54%		1+/41 2%+	19/41 46%+	23+/23 100%+ In another group, 40% on Rx	23 +/ 23 100%+	No value if residual present Not definitive if on Rx
Bayer & McDougall[36]	"Adequate"	<2	(3 assays) 2.5–5	>2.5 –5	15–45% (Tg detectable)						No statement
Ericsson et al[40]	Yes	<3	2 (exclude ab+)	>10	19+/101 19%		1+/80 1%		100%		Tg can replace scan in most cases
Hufner et al[39]	300 μg qd	No data	5	>10					Group 1 old mets 12–/68 18%– Group 2 new mets 2–/20 10%–	22+/28 78%+	Omit if prior scan negative and Tg <5
Valimaki & Lamberg[42]	Yes	No data	3	>3	10+/28 36%+						Use scan only if Tg+
Pacini et al[41]	TRH response negative	No data	3 (exclude) TGHA+)		7%+	45%–			14%–	100%+	Negative assay on Rx does not exclude disease

continued

TABLE 6.2. Selected results of Tg assays in thyroid cancer patients *(continued)*.

Authors and reference	T_4Rx	TSH level (μU/ml)	Assay sensitivity (ng/ml)	Assay +	Patients with presumed normal residual thyroid and no known carcinoma		Patients with *no* residual thyroid and status unknown		Patients with definite carcinoma		Comment
					On T_4	Off T_4	On T_4	Off T_4	On T_4	Off T_4	
Black et al[33]	Yes	No data	No data	>5					14%−		Use Tg rather than scan
Charles et al[34]	0.15–0.3 mg	Negative TRH response	0.6	>7	7+/16 44%+		100%−		8+/8 100%+		Tg can replace scan
Echenique et al[35]	No		%	>20		4+/21 19%+				4−/18 22%−	Use caution; Tg scan complement
Barsano et al[37]	Variable	No data	2–4	>10		41%		3−/9 33%−			Can reduce need for scanning Best off T_4
Feldt-Rasmussen et al[38]	Yes	No data	<5	> Normal or >10			7+/42		14+/14		"Important adjunct"
Girelli et al[43]	Yes	No data	No data	Not stated	24% raised 76% "low"		7.8% raised 98.2% "low"		29% "low"		"A general guide" "Measure off" "Suppressive Rx"
Ashcraft & Van Herle[32]	0.15–0.3 mg	?	1	>1	6+/11 (54%)		15+/18 (83%)		3−/19 (16% low)		Tg <1 on Rx suggests avoid scan

residual thyroid tissue can currently be managed by the criteria given without scans, since a large proportion would have nonsuppressed values and therefore would be subjected to scanning.

Patients with no known residual normal thyroid tissue and with status unknown (i.e., whether metastases were or were not present) were much fewer in number in these studies. In this group the effectiveness of thyroid hormone replacement in suppressing Tg to the acceptable range was often greater with almost all patients being adequately "suppressed." From 1% to 83% had values above the author's "suppressed" level. It is not clear how these patients would fare following the prescription given by Dr. Dunn, since the cutoff point in these patients ranged from 2.5 to 7 ng/ml.

In the patients with definite cancer, there is a great variation in the results. In several studies, 10% to 29% are acceptably negative on suppressive therapy. In other studies, 100% are positive.

The large majority of patients with differentiated thyroid carcinoma appear to be free of malignancy after successful surgery, often including node resection, and often followed by destruction of residual thyroid tissue using radioactive iodide. A small group of patients have known metastatic thyroid carcinoma, and in these patients, the value of the Tg is relatively limited except as a measure of disease progress, since the important therapeutic question is whether radioactive iodide treatment can be used successfully. Thus the important role of Tg in determination of disease status is limited to a small fraction of treated patients, probably not more than 5% or 10%, who have or will develop recurrent thyroid malignancy. Although it would be extremely gratifying to give up radioisotope scans and save our patients the trouble and possible health disadvantages of this procedure, it is not obvious to us that the data presented warrant this position. Clearly, reliance on Tg assay is not possible in patients who have residual normal thyroid tissue in the neck. Very limited data are available using the criteria suggested. A small but significant fraction of patients with carcinoma are not detected by Tg. Data are not available to allow comparison of a group managed prospectively over a long period of time by serial serum Tg assays, versus infrequent but periodic radioactive iodide scanning. Although we may be in a position to switch to the use of Tg with complete reliability in the future, it is not obvious that that time has been reached at present.

Acknowledgment. This commentary was supported by the David Wiener Research Fund.

Response by Dr. John T. Dunn

I was careful to point out the conditions under which I believe it reasonable to omit hypothyroidism and scanning. These included a serum Tg of less than 1 ng/ml in a reliable assay and no special clinical concerns over recurrence. I do not find anything in Dr. DeGroot and Dr. Jahazi's review to refute this recommendation.

I emphasize the importance of a sensitive assay. The studies using the most sensitive assay support my conclusion, as detailed in the paper. Three of the studies (not one, as suggested by Drs. DeGroot and Jahazi), used an assay sensitive to this level.[32-34] A commercial assay (Nichols) is sensitive to 1 ng/ml, so it is within the easy reach of all clinicians.

I agree that by this approach only a fraction of patients with residual thyroid tissue will escape thyroxine withdrawal and scanning. For this fraction, at least, this approach will have clear benefit. Patients with a detectable Tg, and thus requiring hypothyroidism and scanning, will presumably be treated with radioiodine and have the residual thyroid destroyed. After that, in the absence of recurrent tumor, their chances of having a suppressed Tg while on thyroxine are excellent, and they can be managed by following the serum Tg without the hypothyroidism and scanning. As to "how the patients would fare," my answer is that they will probably fare quite well and many can be spared unnecessary hypothyroidism and scanning.

The value of Tg as a measure of disease progress is not to be dismissed lightly. At the least, an elevated or detectable value encourages a vigorous search for functioning metastases and offers the opportunity for earlier detection and treatment. Many published studies cite examples in which the serum Tg was the earliest indicator of persistent or recurring cancer.

Several final points deserve emphasis. First, most patients are profoundly uncomfortable with the periodic hypothyroidism necessary for scanning. In addition to discomfort, the increased TSH carries the risk of promoting tumor growth. Thus, scanning "just to be safe" is unsound. Second, it has still not been convincingly shown that radioiodine therapy—the whole reason for scanning—decreases mortality from papillary cancer. Therefore, the ultimate benefit of hypothyroidism and scanning needs to be weighed against its discomforts and danger. Third, let me reemphasize that my recommendations were offered with the proviso that they be considered in the context of the individual patient. In a patient with a history of resected papillary cancer, on thyroxine therapy, with no special risk factors, and a serum Tg less than 1 ng/ml, I still conclude that withholding thyroxine and scanning are unnecessary and undesirable. My clinical experience with such patients has reinforced this view. Finally, Drs. DeGroot and Jahazi and I fully agree on at least one issue, that more extensive studies are needed to refine our approaches to thyroid cancer. I eagerly await such investigations.

Discussion by Dr. Joel I. Hamburger

How does the Tg compare to whole body imaging in the detection of residual thyroid cancer? At what level of elevation of Tg would whole body imaging be indicated?

The Tg assay has been touted as useful to differentiate factitious thyrotoxicosis (when the level should be low) from the spontaneously resolving hyperthyroidism of thyroiditis (when the level is high). In both situations the RAIU is suppressed. However, if the patient with factitious thyrotoxicosis were taking desiccated thyroid, the Tg might still be elevated. At least this was the case for few patients I have seen. Do you have any experience on this point?

Response by Dr. John T. Dunn

From my review, the serum Tg is simpler, cheaper, and more sensitive in recognizing recurrence of differentiated thyroid cancer than is the whole body scan. Therefore, if the serum Tg is suppressed or is undetectable, I would not consider further imaging necessary.

As regards the magnitude of elevation at which imaging would be considered necessary, this is a difficult question to answer. If a patient has had a total thyroid ablation and still has detectable serum thyroglobulin, we need to worry about the continued presence of thyroid tissue and/or cancer, and imaging will probably be necessary.

In regard to whether Tg is elevated if the patient takes desiccated thyroid in thyrotoxicosis factitia, I have had no personal experience with this. I am interested in the patients you describe. I can speculate that since desiccated thyroid is largely thyroglobulin, it might still retain enough antigenic sites to be recognized by anti-Tg antibodies, particularly if it is not completely broken down in the GI system before absorption.

Discussion by Dr. Ian D. Hay

Can serum Tg determinations be used to monitor patients with thyroid cancer who have undergone less than total thyroidectomy but are on T_4 suppressive therapy? Is it likely that assays can be developed that can accurately measure serum Tg levels despite the presence of anti-Tg autoantibodies? If a patient's serum Tg is <1 ng/ml on T_4 suppressive therapy, and the patient is considered at high risk for recurrence, should whole body ^{131}I scanning be performed at regular (? every 3 to 5 years) time intervals during follow-up? Is there a future for radioiodine-labeled anti-Tg antibodies in the therapy of metastatic differentiated thyroid cancer?

Response by Dr. John T. Dunn

Can serum Tg determinations be used to monitor patients with thyroid cancer who have undergone less than total thyroidectomy but are on T_4 suppression therapy? The serum Tg determination is less useful in these patients. The problem is that remnants of normal thyroid may continue to secrete Tg even when on T_4 therapy. The Tg will be useful if undetectable, in which case residual cancer is unlikely, and also if elevated, which raises a strong suspicion of persistent cancer. However, if the Tg is in the normal range, you cannot tell whether this Tg is secreted from the remnant of normal thyroid or by remaining thyroid cancer.

Is it likely that assays can be developed that can accurately measure serum Tg levels despite the presence of anti-Tg autoantibodies? The answer is yes. Black et al[33] reported that their particular antiserum was not influenced by anti-Tg autoantibodies, suggesting it recognized different antigenic sites. Experience with monoclonal antibodies shows a number of different epitopes in Tg, and it is reasonable to expect that selected monoclonals can react specifically with antigenic sites not attacked by autoantibodies. Izumi et al[63] recently reported such a monoclonal antibody.

If a patient's serum Tg is less than 1 ng/ml on T_4 suppressive therapy, and the patient is considered at high risk for recurrence, should whole body ^{131}I scanning be performed at regular (? every 3 to 5 years) time intervals during follow-up? This decision would have to be individualized to the particular patient, particularly as to what constitutes the high risk. The general experience, as detailed in my review, is that the serum thyroglobulin predicts metastases or recurrent disease better than does the whole body scan, especially in the studies with assay sensitivity to 1 ng/ml. In the patient described I would be content not to repeat the thyroid scan every 3 to 5 years, but I emphasize that this decision would be heavily influenced by general clinical considerations and by my level of concern about the risk of recurrence.

Is there a future for radioiodine-labeled anti-Tg antibodies in the therapy of metastatic differentiated cancer? This is an interesting possibility, although I do not know of current work on this. The same objective is probably achieved with our current and much simpler approach, the direct administration of radioiodine, since ^{131}I will go to differentiated thyroid cancer and essentially nowhere else. There is a considerable parallel between the tumor's secretion of thyroglobulin and its radioiodine uptake; thus the tumor secreting significant amounts of Tg, and therefore a potential candidate for antibody therapy, could also be treated directly with radioiodine.

Discussion by Dr. Leonard Wartofsky

Are there any useful techniques for correction of Tg RIA values when antibodies are present in the sera, such as measuring the percent binding on the original and all subsequent samples in order to facilitate following such patients to detect serial increases in Tg levels relative to the original baseline value? Are there any advantages to assay of Tg by IRMA rather than RIA? Do you think that the American Thyroid Association (ATA) should take a position on the use and utility of Tg assay in the follow-up of thyroid cancer?

Response by Dr. John T. Dunn

I do not know any current practical techniques for correction of the Tg RIA values when autoantibodies are present. Your suggestion about measuring the percent binding on the original and subsequent samples would probably depend on a constant serum level of autoantibodies over time, and I am not sure we could expect that. As stated above, the best approach will probably be developing antibodies without cross-reactivity.

Regarding the advantages of an IRMA rather than RIA, I would think the main point again would be the ultimate sensitivity of the assay. Of the studies I reviewed, only that by Pacini et al[41] used an IRMA and in that the sensitivity was not quite as good as that of the RIA used by Van Herle et al.[19] In such an instance I would favor the RIA.

Regarding whether the ATA should take a position on the use of Tg in the follow-up of thyroid cancer patients, I believe conclusions are still too tentative to be an official policy. It is certainly justifiable to point out that the serum Tg is a useful marker for differentiated thyroid cancer and that it may be more reliable than the whole body scan. We still need more experience before we can conclude with certainty that a suppressed serum Tg while on thyroxine therapy is satisfactory insurance against the continuing presence of cancer. Also, having failed several years ago to persuade the ATA to adopt a resolution urging the eradication of endemic goiter in developing countries, I think it would be difficult to get something more controversial, such as this, through that body.

Discussion by Dr. John E. Freitas

My dissatisfaction with serum thyroglobulin in the monitoring of thyroid cancer patients most likely ensues from our use of an insensitive assay similar to that employed by Echenique et al.[35] Our comparison of this serum thyroglobulin assay versus ^{131}I whole body scanning demonstrated a 36% discordance. Regional lymph node metastases commonly exhibited serum thyroglobulin values $<$20 ng/ml despite positive ^{131}I WBS.

Discussion by Dr. James C. Sisson

The measurement of thyroglobulin is an important advance in the staging of well-differentiated thyroid cancer. But we must ask an additional question: how reliable are the changes in thyroglobulin levels in predicting recurrence or progression of thyroid cancer? For most patients with thyroid cancer, the stage of disease after initial treatments (thyroidectomy and possibly ^{131}I) will be determined by several modalities that will include scintigraphy and the thyroglobulin concentration. When satisfied with the results of the treatments, the clinician must adopt a strategy that will tell him or her if and when further treatment is necessary. As a major indicator of recurrent cancer, thyroglobulin levels must then be defined in terms of normal variability from year to year. It may be that a persistently undetectable concentration of thyroglobulin will be sufficient evidence for absence of cancer. But the thyroglobulin levels in some patients without cancer may not remain imperceptible. Particularly, patients may retain small amounts of normal thyroid tissue that are not worthy of ^{131}I treatment, and, as a consequence, low levels of serum thyroglobulin will be found in their sera. Also, in some patients multiple treatments with ^{131}I may leave residual tumor, which at the time appears stable. What change or what pattern of change in thyroglobulin levels predicts recurrence or progression of thyroid cancer?

Response by Dr. John T. Dunn

I would be concerned about a recurrence if a patient initially has a suppressed Tg level that subsequently becomes detectable or elevated. Many patients with progressive thyroid cancer and enlarging metastases will increase the serum Tg in parallel to the increase in tumor size. An increasing serum Tg should give real concern about tumor progression.

Commentary by Dr. Martin J. Schlumberger and Dr. Maurice R. Tubiana: Serum Tg Measurements and Total Body ^{131}I Scans in the Follow-up of Thyroid Cancer Patients

Thyroglobulin (Tg) has been shown to be a reliable marker for the follow-up of patients with differentiated thyroid carcinoma. Discrepancies between serum Tg levels and other tests indicating the presence of neoplastic tissue are rare.

The aim of this commentary is to review briefly the main data published since the availability of Tg assay and, in particular, results obtained since 1977 at Institut Gustave-Roussy (IGR), Villejuif. From these data a practical scheme for the combined use of serum Tg measurement and total body ^{131}I scan has been developed.

Methods of Measurement. In most studies concerning the measurement of Tg, including those carried out at Institut Gustave-Roussy,[45] a double-antibody radioimmunoassay was used and derived from the original technique of Van Herle et al.[28] Immunoradiometric assays[46,47] and ELISA[48] have also been described. Quality controls of each of these assays are satisfactory. However, wide variations obtained both in normal persons and in patients with differentiated thyroid carcinoma have been assumed to derive, at least in part, from variations in assays used. Two recent studies[29,49] have clearly demonstrated that results obtained with given sera depend upon the components used for each specific technique. For these reasons, a European cooperative trial is preparing a stable and reproducible Tg calibrator.[49]

When used in the follow-up of patients with differentiated thyroid carcinoma, the assay has to be sensitive enough to detect concentrations as low as 1 to 2 ng/ml.[50] In fact, most commercial kits can only detect 3 to 5 ng/ml. Furthermore, the smallest detectable doses are calculated according to classical methods,[51] and are often higher with the routine use of these assays. Recently, an immunoradiometric assay using two different monoclonal antibodies was set up at the Institut Gustave-Roussy, Tg concentrations as low as 0.5 ng/ml have been detected, and results are strongly correlated with those of the RIA in patients without anti-Tg autoantibodies.

Endogenous anti-Tg antibodies are found in 1% to 40% of patients with differentiated thyroid carcinoma.[50] They interfere in most assays and prevent the determination of the actual Tg level.[49] Thus, endogenous anti-Tg antibodies invariably lead to decreased values or even to false negative results with immunoradiometric assays,[47] even if monoclonals are used. In radioimmunoassays, they lead to either falsely depressed or elevated Tg values that mainly depend on cross-reactivity of the second antibody with endogenous human IgG and on the ratio of antigen to antibody.[38,52] Hence, for most authors, their presence precludes any clinical interpretation of the results, and techniques used for their detection should be sensitive enough to detect all antibodies that may interfere in the assay. However, in one assay, results obtained in these patients were in accordance with radiologic and isotopic findings.[53] Moreover, it is possible that monoclonal antibodies reacting with epitopes not recognized by autoantibodies may be a clue for measuring Tg in the presence of autoantibodies.

Many studies have indicated that TSH is one of the regulators of Tg release by normal thyroid tissues and by goiters.[25] Most metastases of differentiated thyroid carcinoma are TSH responsive, even if they do not pick up radioiodine.[54,55] After T_3 withdrawal, in a given patient, there is a parallel increase in serum TSH and Tg levels, until a plateau is reached, some 15 days after T_3 withdrawal.[54] Furthermore, during T_4 treatment, Tg level is strongly dependent upon TSH level. The decrease of TSH from normal range to low or undetectable level, as shown by sensitive techniques, induces wide variations in serum Tg level. Slight variations in T_4 daily dose, even where TSH remains undetectable, induce significant changes in serum Tg level.[56] Therefore, in interpreting Tg measurements, conditions (on or off T_4 treatment) and TSH level measured by a sensitive technique (low or undetectable) have to be taken into account.

Results in Cancer Patients: Patients with Metastases. Serum Tg level is elevated in most patients with metastases of differentiated thyroid carcinoma,[31,50] whatever the age of the patient, the histologic type of the thyroid tumor, or radioiodine uptake.[57]

False-negative measurements account for 0% to 10% of the patients with known distant metastases,[50] and are even more infrequent when sensitive assays are used. AT IGR, serum Tg level measured at the time of the detection of distant metastases in 126 patients during T_4 treatment was only undetectable in one patient. This patient had lung metastases that were not visible on chest X ray yet were detected by a total body scan performed with 1 mCi of ^{131}I, 6 months after initial treatment. Tg level was detectable in all the other patients, and was above 10 ng/ml in 110. Fifteen patients had a detectable but relatively low Tg level (< 10 ng/ml), nine of whom had lung metastases shown only by a total body scan, and not visible on a chest X ray. Tg was frequently undetectable during T_4 treatment in patients with small lymph nodes, which were found to be histologically involved.

A recent report concerns 16 patients with an undetectable Tg level in spite of the presence of lymph node involvement revealed by sonography or total body ^{131}I scan.[58] In our series, of 14 patients with lymph node involvement, Tg level was undetectable in two, below 10 ng/ml in eight, and above 10 ng/ml in only four patients. This underlines the need for systematic lymph node dissection at initial treatment, and thereafter a systematic total body ^{131}I scan to confirm the absence of any ectopic uptake in the neck.

TSH stimulation increases the sensitivity of Tg measurement. Nevertheless, false negative Tg measurements have been reported in this situation in a few patients with metastatic spread. However, technical problems in Tg assays and negative criteria, which varied considerably among reports, may partially explain these discrepant results. At IGR, Tg level measured after thyroid hormone withdrawal was above 10 ng/ml in all the 152 patients with distant metastases and above 40 ng/ml in 138 of them.

The precise amount of neoplastic tissue necessary to increase serum Tg level is unknown. Tg level was frankly elevated in patients with multiple large metastases, it frequently remained relatively low when metastases were not visible on X rays, and was low or even undetectable in patients with small lymph node metastases (Fig. 6.1). This suggests that there is a rough correlation between the level of the serum Tg and the tumor burden. We have shown that the size of the metastases at their discovery had a paramount prognostic importance both with respect to achievement of complete remission and survival.[57] Tg measurement must therefore be adequately sensitive to reliably detect these low Tg levels, which warrant further investigation.

Results in Cancer Patients: No Evidence of Disease. Tg level during T_4 treatment was undetectable in more than 80% of the patients without thyroid remnants and no evidence of disease.[50] At IGR, Tg level was undetectable in 87% and below 10 ng/ml in 98%. Following thyroid hor-

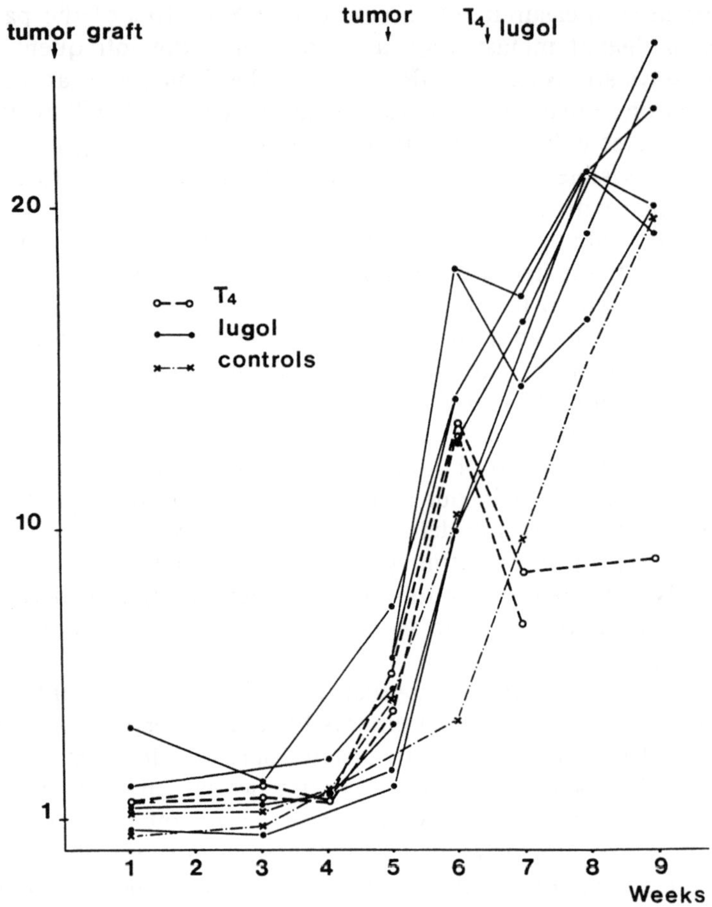

FIGURE 6.1. Tg level in rats following graft of 1–1C2 rat transplantable thyroid tumor. Tg level increased before tumors became palpable and thereafter was related to tumor size.

mone withdrawal, Tg level was detectable in 49% of these patients but remained below 10 ng/ml in 85%. Therefore, Tg level measurement after thyroid hormone withdrawal is particularly useful in patients with detectable but low (< 10 ng/ml) Tg level during T_4 treatment. In 85% of the patients with no other evidence of disease, it remained below 10 ng/ml, whereas in all patients with distant metastases it increased to above 10 ng/ml, and was above 40 ng/ml in 91% of them.

We have shown interest in the administration of 100 mCi of ^{131}I for the detection of unknown metastases in high-risk patients.[59] This method is particularly useful in patients with detectable Tg levels during T_4 treatment. At IGR, in 22 patients with elevated Tg levels (> 10 ng/ml during

T_4 treatment and > 40 ng/ml following thyroid hormone withdrawal), and no clinical and radiologic evidence of the disease, total body ^{131}I scans performed after the administration of 100 mCi showed distant metastases in eight, neck recurrence in four, and normal thyroid remnants in six. Furthermore, it allowed the demonstration of lung uptake in one third of our patients with lung metastases and normal chest X ray.[57] In a recent study, among 135 patients, 17 had detectable Tg levels and negative total body ^{131}I scans performed with 5 mCi. A total body scan performed after the administration of 100 mCi showed significant residual or metastatic uptake in all patients but one.[60] Due to the slow growth rate of thyroid cancer, Tg level may be elevated in these patients for years or even decades before neoplastic tissue becomes detectable on X rays.

In patients with thyroid remnants, although results were less satisfactory, Tg measurement is useful for detecting relapse.[61] During T_4 treatment, Tg was undetectable in only one third of the patients without evidence of disease and below 10 ng/ml in 73%. A log rank test has shown that in these patients, the risk of relapse was low in patients whose Tg levels were below 10 ng/ml during T_4 treatment. Above 10 ng/ml, the higher the Tg level, the greater the risk of detectable disease. However, the interpretation of the increase of Tg level following TSH stimulation in these patients is more difficult since Tg can be released either by normal thyroid remnants or by neoplastic foci.

Strategy of Follow-up. A strategy has been developed for the follow-up of patients without anti-Tg autoantibodies. During T_4 treatment, Tg level is taken into account only when TSH level is below 0.1 μU/ml. When TSH is above 0.1 μU/ml, the daily dose of T_4 is increased by 25 μg, and T_4, TSH, and Tg levels are remeasured 3 months later.[62] It should be mentioned that establishing threshold values may be valid only for a specific Tg assay because Tg assays still lack standardization.

As shown previously, the ablation of thyroid remnants increases the sensitivity of both Tg measurement and total body ^{131}I scan and is advocated in patients with high risk of developing relapses.

In patients without thyroid remnant, a total body ^{131}I scan is performed each year for 2 or 3 years after initial therapy. Thereafter, in patients with undetectable Tg levels or levels below 10 ng/ml after thyroid hormone withdrawal, follow-up is resumed with clinical evaluation, T_4, TSH, and Tg measurements, once a year, while on T_4 therapy. Since it was reported that an undetectable Tg level might not totally preclude a relapse, a prospective study was carried out, including a total body ^{131}I scan and a chest X ray every 5 years. By using this protocol in 500 patients since 1977, we have not found any metastases in patients where Tg levels were below 10 ng/ml after thyroid hormone withdrawal. This suggests that, in these patients, it is possible to avoid total body ^{131}I scans for at least 10 years. When Tg level is detectable during T_4 treatment, a total body ^{131}I scan and chest X ray should be performed. Patients with positive total body scans are treated with 100 mCi of ^{131}I. In those patients with a negative total body ^{131}I scan, where Tg levels are higher than 10 ng/ml after thyroid hormone withdrawal, additional attempts to localize thyroid tissue should be made (neck ultrasound, neck and thorax CT scan). A dose of 100 mCi of ^{131}I is administered when Tg level increases to above 40 ng/ml following thyroid hormone withdrawal. In fact, in most of these patients, normal or neoplastic tissue was demonstrated by a posttherapy scan[60] at the time of Tg measurement. However, in some patients, all these attempts remain negative and it seems advisable to perform them again within 1 or 2 years. Eight of such patients were followed at IGR for more than 4 years. In two, Tg level increased with time, and lung metastases without radioiodine uptake were discovered within 2 years after the first Tg measurement (Fig. 6.2). In the other six patients, Tg level remained stable with time and no other evidence of disease was discovered. This may be related to the slow growth rate of thyroid tumors.

Patients with thyroid remnants, whose Tg levels are higher than 30 ng/ml during T_4 treatment, are given 100 mCi of radioiodine with a total

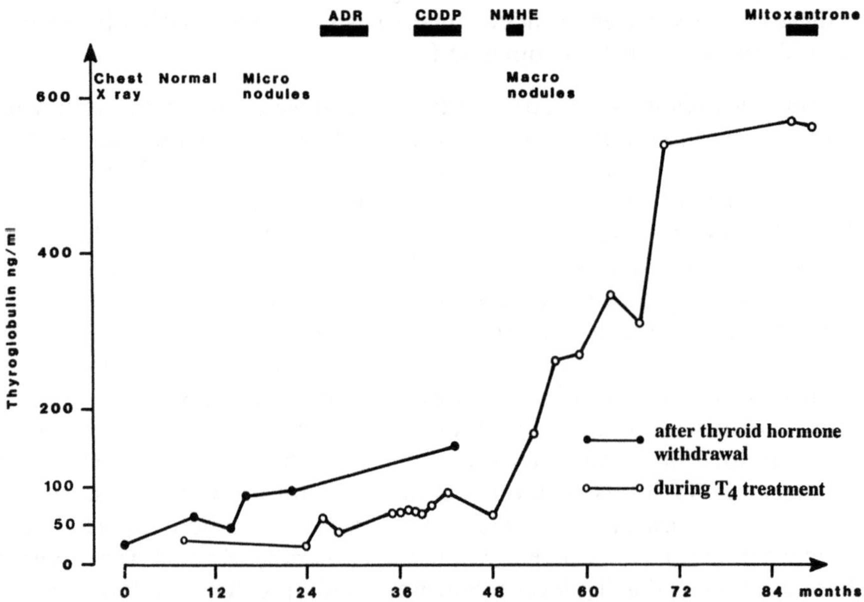

FIGURE 6.2. Tg level in a patient aged 52 years, treated for a moderately differentiated follicular carcinoma. Tg was detectable before lung metastases were visible on chest X ray. No uptake was found in the metastases at a stage at which they were macronodular and even following 100 mCi of radioiodine. Note that there was apparently no effect of chemotherapy. ADR = adriamycine; CDDP = cis Pt; NMHE = ellipticine.

body scan 5 days later. When Tg levels are detectable but low (< 10 ng/ml), they are followed like those with undetectable Tg levels. When Tg level ranges between 10 and 30 ng/ml, a total body ^{131}I scan and a chest X ray are performed, especially if Tg level increased at consecutive determinations.

Since the routine use of the TG assay, the number of total body ^{131}I scans performed has been reduced but they are more effective because their indications are better defined. Before 1977, 13% of our patients with lung metastases had a normal chest X ray, whereas after the introduction of serum Tg as part of the follow-up, the proportion of patients with normal X rays at the time of metastasis detection rose to 43%.[57] This earlier discovery of relapses should improve the long-term treatment results (Fig. 6.3).[57]

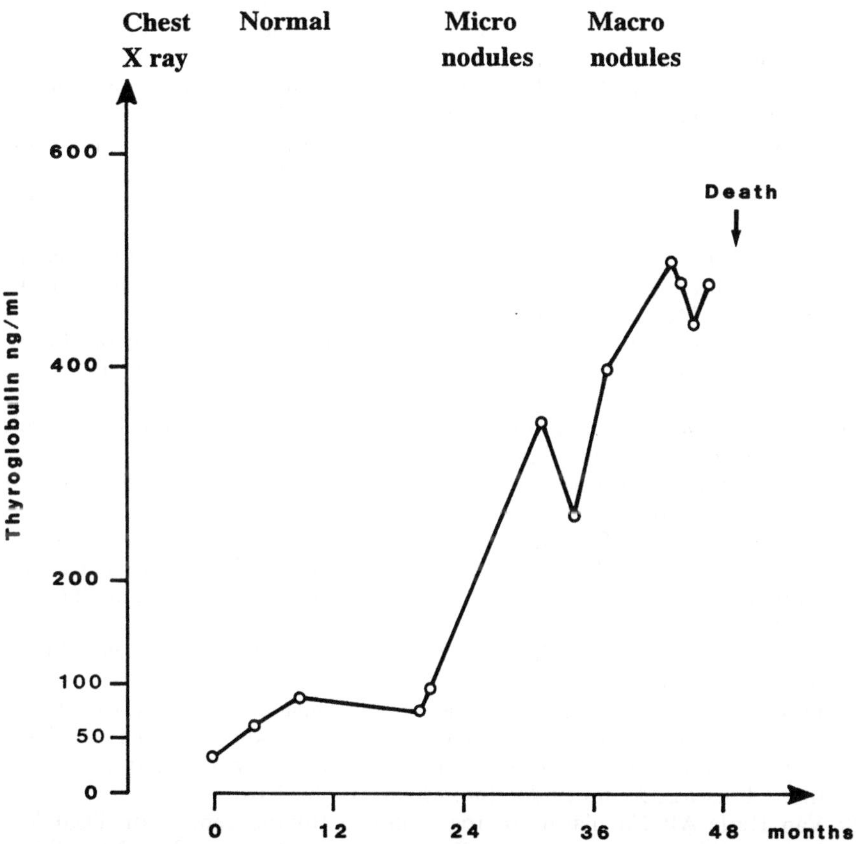

FIGURE 6.3. Tg level in a patient aged 41 years, treated for a papillary carcinoma. Tg was detectable under thyroxine (T_4) treatment, at least 2 years before metastases became detectable on chest X rays. No uptake was found in lung metastases.

References

1. Eggo MC, Burrow GN (eds): Progress in Endocrine Research and Therapy: vol. 2. Thyroglobulin—The Prothyroid Hormone. New York: Raven Press, 1985.
2. Dunn JT: Thyroglobulin: Structure, function and clinical relevance. Thyroid Today 1985;8:1.
3. Dunn JT, Ray SC: Changes in thyroglobulin structure after TSH administration. J Biol Chem 1975;250:5801.
4. Rawitch AB, Gregg J, Turner DC: Nature and location of hormonogenic sites within the structure of thyroglobulin. In: Eggo MC, Burrow GN, eds. Thyroglobulin—The Prothyroid Hormone. New York: Raven Press, 1985, p 43.
5. Marriq C, Lejeune PJ, Rolland M, Lissitzky S: Structure of thyroid hormone-containing peptides in porcine and human thyroglobulins. In: Eggo MC, Burrow GN, eds. Thyroglobulin—The Prothyroid Hormone. New York: Raven Press, 1985, p 21.
6. Dunn JT, Anderson PC, Fox JW, et al: The sites of thyroid hormone formation in rabbit thyroglobulin. J Biol Chem 1987;262:16948.
7. Lamas L, Anderson PA, Fox JW, Dunn JT: Early iodination sites of thyroglobulin. Ann D'Endocrinologie 1987;48:abstr 62.
8. Fassler CA, Dunn JT, Anderson PC, et al: TSH alters the utilization of thyroglobulins hormonogenic sites. J Biol Chem 1988;263:17366.
9. Dunn JT, Ray SC: Variations in structure of thyroglobulins from normal and goitrous human thyroids. J Clin Endocrinol Metab 1978;47:861.
10. Cave WT Jr, Dunn JT: Studies on the thyroidal defect in an atypical form of Pendred's syndrome. J Clin Endocrinol Metab 1975;41:590.
11. Dunn JR, Ray SC: Changes in iodine metabolism and thyroglobulin structure in metastatic follicular carcinoma of the thyroid with hyperthyroidism. J Clin Endocrinol Metab 1973;36:1088.
12. Silva JE, Santelices R, Kishihara M, Schneider A: Low molecular weight thyroglobulin leading to a goiter in a 12-year-old girl. J Clin Endocrinol Metab 1984;58:526.
13. Lever EG, Medeiros-Neto GA, DeGroot LJ: Inherited disorders of thyroid metabolism. Endocrinol Rev 1983;4:213.
14. Byfield PGH, Clingan D, Himsworth RL: Exposure of thyroxine residues in human thyroglobulin. Biochem J 1984;219:405.
15. Kohno Y, Tarutani O, Sakata S, Nakajima H: Monoclonal antibodies to thyroglobulin elucidate differences in protein structure of thyroglobulin in healthy individuals and those with papillary adenocarcinoma. J Clin Endocrinol Metab 1985;61:343.
16. Kurata A, Ohta K, Mine M, et al: Monoclonal antihuman thyroglobulin antibodies. J Clin Endocrinol Metab 1984;59:573.
17. DeMicco D, Ruf J, Carayon P, et al: Immunohistochemical study of thyroglobulin in thyroid carcinomas with monoclonal antibodies. Cancer 1987;59:471.
18. Kim PS, Dunn AD, Dunn JT: Altered immunoreactivity of thyroglobulin in thyroid disease. J Clin Endocrinol Metab, 1988;67:161.
19. Van Herle AJ: Circulating thyroglobulin in thyroid cancer. In: Eggo MC, Burrow GN, eds. Thyroglobulin—The Prothyroid Hormone. New York: Raven Press, 1985, p 317.
20. Schneider AB: Physical characterization of serum thyroglobulin: Implications for the mechanism of secretion into the circulation. In: Eggo MC, Burrow

GN, eds. Thyroglobulin—The Prothyroid Hormone. New York: Raven Press, 1985, p 297.
21. Pinchera A, Pacini F, Mariotti S, Martino E: Recent studies on the application of serum thyroglobulin measurement in thyroid disease. In: Eggo MC, Burrow GN, eds. Thyroglobulin—The Prothyroid Hormone. New York: Raven Press, 1985, p 307.
22. Herzog V: Vesicular transport of thyroglobulin across the follicular wall. In: Eggo MC, Burrow GN, eds. Thyroglobulin—The Prothyroid Hormone. New York: Raven Press, 1985, p 143.
23. Ikekubo K, Kishihara M, Sanders J, et al: Differences between circulating and tissue thyroglobulin in rats. Endocrinology 1981;109:427.
24. Unger J, Van Heuverswyn B, Decoster C, et al: Thyroglobulin and thyroid hormone release after intravenous administration of bovine thyrotropin in man. J Clin Endocrinol Metab 1980;51:590.
25. Uller RP, Van Herle AJ, Chopra IJ: Comparison of alterations in circulating thyroglobulin, triiodothyronine and thyroxine in response to exogenous (bovine) and endogenous (human) thyrotropin. J Clin Endocrinol Metab 1973;37:741.
26. Schneider AB, Fleischmann K, Chu L: Thyrotropin increases the iodine content of rat circulating thyroglobulin as measured by equilibrium density gradient centrifugation. Biochem Biophys Acta 1985;838:329.
27. Gebel F, Ramelli R, Burgi U, et al: The site of leakage of intrafollicular thyroglobulin into the bloodstream in simple human goiter. J Clin Endocrinol Metab 1983;57:915.
28. Van Herle AJ, Uller RP, Matthews NL, Brown J: Radioimmunoassay for measurement of thyroglobulin in human serum. J Clin Invest 1973;52:1320.
29. Van Herle AJ, Van Herle IS, Greipel MA: An international cooperative study evaluating serum thyroglobulin standards. J Clin Endocrinol Metab 1985;60:338.
30. Van Herle AJ: Serum thyroglobulin measurement in the diagnosis and management of thyroid disease. Thyroid Today 1981;4.
31. Van Herle AJ, Uller RP: Elevated Serum Thyroglobulin. J Clin Invest 1975;56:272.
32. Ashcraft MW, Van Herle AJ: The comparative value of serum thyroglobulin measurements and iodine 131 total body scans in the follow-up study of patients with treated differentiated thyroid cancer. Am J Med 1981;71:806.
33. Black EG, Cassoni A, Gimlette TMD, et al: Serum thyroglobulin in thyroid cancer. Lancet 1981;2:443.
34. Charles MA, Dodson LE JR, Waldeck N, et al: Serum thyroglobulin levels predict total body iodine scan findings in patients with treated well-differentiated thyroid carcinoma. Am J Med 1980;69:401.
35. Echenique RL, Kasi L, Haynie TP, et al: Critical evaluation of serum thyroglobulin levels and ^{131}I scans in post-therapy patients with differentiated thyroid carcinoma: Concise communication. J Nucl Med 1982;23:235.
36. Bayer MR, McDougall IR: Differences in radioimmunoassay results for thyroglobulin that affect management of patients with thyroid cancer. Clin Chem 1984;30:81.
37. Barsano CP, Skosey C, DeGroot LJ, Refetoff S: Serum thyroglobulin in the management of patients with thyroid cancer. Arch Intern Med 1982;142:763.

38. Feldt-Rasmussen U, Holten I, Hansen HS: Influence of thyroid substitution therapy and thyroid autoantibodies on the value of serum thyroglobulin in recurring thyroid cancer. Cancer 1983;51:2240.
39. Hufner M, Stumpf JP, Grussendort M, et al: A comparison of the effectiveness of ^{131}I whole body scans and plasma Tg determinations in the diagnosis of metastatic differentiated carcinoma of the thyroid: A retrospective study. Acta Endocrinol 1983;104:327.
40. Ericsson U-B, Tegler L, Lennquist S, et al: Serum thyroglobulin in differentiated thyroid carcinoma. Acta Chir Scand 1984;150:367.
41. Pacini F, Lari R, Mazzeo S., et al: Diagnostic value of a single serum thyroglobulin determination on and off thyroid suppressive therapy in the follow-up of patients with differentiated thyroid cancer. Clin Endocr 1985;23:405.
42. Valimaki M, Lamberg B-A: How to deal with undetectable and low measurable serum thyroglobulin levels in the follow-up of patients with differentiated thyroid carcinoma? Acta Endocrinol 1985;110:487.
43. Girelli ME, Busnardo B, Amerio R, et al: Critical evaluation of serum thyroglobulin (Tg) levels during thyroid hormone suppressive therapy versus Tg levels after hormone withdrawal and total body scan: Results in 291 patients with thyroid cancer. Eur J Nucl Med 1986;11:333.
44. Schlumberger M, Fragu P, Paramentier C, Tubiana M. Thyroglobulin assay in the follow-up of patients with differentiated thyroid carcinomas: Comparison of its values in patients with or without normal residual tissue. Acta Endocrinol 1981;98:1.
45. Schlumberger M, Fragu P, Vignal A, Tubiana M: Dosage de la thyroglobuline humaine dans le serum et le tissu thyroidien. Ann Endocrinol (Paris) 1979;40:439.
46. Bayer MF, Kriss JP: Immunoradiometric assay for serum thyroglobulin: Semi-quantitative measurement of thyroglobulin in antithyroglobulin positive sera. J Clin Endocrinol Metab 1979;49:557.
47. Mariotti S, Cupini C, Ciani C, et al: Evaluation of a solid-phase immunoradiometric assay (IRMA) for serum thyroglobulin: Effect of anti-thyroglobulin auto-antibody. Clin Chim Acta 1982;123:347.
48. Endo Y, Nakano J, Ohtaki S, et al: An enzyme immunoassay for the measurement of thyroglobulin in human serum. Clin Chim Acta 1979;95:325.
49. Feldt-Rasmussen U, Schlumberger M: European interlaboratory comparison of serum thyroglobulin measurement. J Endocrinol Invest 1988;11:175.
50. Hufner M, Reiners C (eds): Thyroglobulin and Thyroglobulin Antibodies in the follow-up of Thyroid Cancer and Endemic Goiter. Stuttgart: Georg Thieme Verlag, 1987.
51. Rodbard D, Rayford PL, Cooper JA, Ross GT: Statistical quality control of radioimmunoassay. J Clin Endocrinol Metab 1968;28:1412.
52. Schneider AB, Pervos R: Radioimmunoassay of human thyroglobulin: Effect of antithyroglobulin auto-antibodies. J Clin Endocrinol Metab 1978;47:126.
53. Tang Fui SC, Hoffenberg R, Maisey MN, Black EG: Serum thyroglobulin concentrations and whole-body radioiodine scan in follow-up of differentiated thyroid cancer after thyroid ablation. Br Med J 1979;2:298.
54. Schlumberger M, Charbord P, Fragu P, et al: Circulating thyroglobulin and thyroid hormones in patients with metastases of differentiated thyroid carcinoma: Relationship to serum thyrotropin levels. J Clin Endocrinol Metab 1980;51:513.

55. Schneider AB, Line BR, Goldman JM, Robbins J: Sequential serum thyroglobulin determinations ^{131}I scans and ^{131}I uptakes after triiodothyronine withdrawal in patients with thyroid cancer. J Clin Endocrinol Metab 1981;53:1199.
56. Spencer CA, Lai Rosenfeld AO, Guttler R: Thyrotropin secretion in thyrotoxic and thyroxine-treated patients: Assessment by a sensitive immunoenzymometric assay. J Clin Endocrinol Metab 1986;63:349.
57. Schlumberger M, Tubiana M, deVathaire F, et al: Long-term results of treatment of 283 patients with lung and bone metastases from differentiated thyroid carcinoma. J Clin Endocrinol Metab 1986;63:960.
58. Muller-Gartner HW, Schneider C: Clinical evaluation of tumor characteristics predisposing serum thyroglobulin to be undetectable in patients with differentiated thyroid cancer. Cancer 1988;61:976.
59. Tubiana M, Mabille JP: L'iode radioactif et le cancer du corps thyroide. I-Problemes diagnostiques. J Radiol (Paris) 1963;44:179.
60. Pacini F, Lippi F, Formica N, et al: Therapeutic doses of iodine-131 reveal undiagnosed metastases in thyroid cancer patients with detectable serum thyroglobulin levels. J Nucl Med 1987;28:1888.
61. deVathaire F, Blanchon S, Schlumberger M: Thyroglobulin level helps to predict recurrence after lobo-isthmusectomy in patients with differentiated thyroid carcinoma. Lancet 1988; i:52.
62. Schlumberger M, de Vathaire F, Wu-Ahouju G, et al: Surveillance post-opêratoire des épithéliomas thyroïdiens différenciés. Apport du dosage ultrasensible de TSH. Press Med 1987;16:1791.
63. Izumi M, Sritongkul N, Kubo I, et al: Measurement of thyroglobulin (Tg) concentrations in sera of patients with anti Tg autoantibody. Abstracts, 69th Annual Meeting, Endocrine Society, June 1987, p. 115.

7

Thyroid Imaging: Indications and Limitations

JOHN E. FREITAS

Over the past 4 decades, there has been considerable progress in radiopharmaceuticals and instrumentation for thyroid imaging that has allowed the imaging physician to provide better anatomic and functional information to his or her referring physicians. Health care cost containment is now a major factor in medical practice. Thyroid imaging procedures are overutilized and considerable cost savings could be achieved by restricting their use to situations in which they would improve patient care.[1,2] Thyroid imaging by radionuclides, computed tomography (CT), ultrasound, X ray fluorescence, and nuclear magnetic resonance (NMR) is possible, but the following review emphasizes the radionuclide approach, which varies throughout the country in both technique and indications.[3,4] The other imaging modalities will be discussed as warranted.

Radiopharmaceuticals and Instrumentation

Tc-99m pertechnetate (Tc-99m) and ^{123}I are used for more than 97% of radionuclide thyroid imaging procedures, but ^{131}I remains useful for certain clinical applications. Tc-99m is preferred by many imaging physicians because of its low cost (<10¢ per study), ready availability, low radiation dose, and patient convenience. During a single patient visit, multiple images (anterior and oblique) of the thyroid can be obtained utilizing a pinhole collimator within 20 minutes of the injection of 5 to 10 mCi of Tc-99m. Images should be correlated with palpatory findings at the time of scanning with the use of various markers. Other physicians prefer ^{123}I because it maps thyroidal organification of iodine whereas Tc-99m is trapped but not organified. However, discordant images between these two tracers are rare (<1% to 2%) and usually associated with benign thyroid disease. ^{123}I thyroid imaging requires two patient visits (4 to 6 or 24 hours between ingestion of 400 μCi and imaging), is more expensive ($60 per dose), is less readily available, and has a similar radiation exposure to Tc-99m. Imaging with a pinhole collimator and appropriate markers is similar to Tc-99m imaging. Many factors such as physician bias, referral patterns, cost, and availability determine which tracer is preferred in a given setting.

Clinical Indications and Limitations

Thyroid imaging provides useful information to the referring physician in the following clinical situations:

1. Functional evaluation of solitary or dominant thyroid nodule
2. Evaluation of upper mediastinal mass
3. Differentiation of Plummer's disease from nodular Graves' disease
4. Functional assessment of thyroid remnant postthyroidectomy for benign disease
5. Detection and staging of residual or metastatic thyroid cancer
6. Evaluation of the cause of neonatal hypothyroidism
7. Evaluation of midline neck masses

Solitary or Dominant Thyroid Nodule

The palpation of a thyroid nodule raises the possibility of thyroid cancer, yet thyroid cancer is uncommon (incidence 10,600 per year) and not readily distinguished from a benign nodule (incidence 300,000 per year).[5–7] Pinhole thyroid imaging with Tc-99m or ^{123}I characterizes nodules as hyperfunctioning, hypofunctioning, or indeterminate. The hyperfunctioning nodule is almost always benign. Discordant imaging (hot on Tc-99m, cold on ^{123}I) is rare and usually associated with benign disease, but thyroid cancer may show this pattern. Hyperfunctioning nodules may or may not be toxic with the former showing suppression of extranodular thyroid tissue. Hyperfunctioning nodules are found in 5% to 15% of patients referred for imaging of solitary or dominant nodules. The hypofunctioning nodule is found in 80% to 90% of solitary nodules and can be caused by multiple pathologic processes. Only 6% to 20% of cold nodules are malignant. An indeterminate nodule found in 4% to 6% of solitary palpable nodules is one that is readily palpable but is not delineated on scan. Such nodules do not distort the thyroid gland outline and do not appear as cold or hot nodules even with oblique angle imaging.[8] Indeterminate nodules have the same significance as cold nodules and the same differential diagnosis. In most instances, these nodules are anterior to functioning thyroid tissue and are not delineated on scan because of the underlying normal activity. These nodules are only discovered if the palpatory findings are correlated with the scan appearance. If hypofunctioning and indeterminate nodules are considered positive findings, the sensitivity of pinhole thyroid imaging in the detection of thyroid cancer should exceed 97%. However, its specificity (15% to 20%) and predictive value of a positive test (16% to 23%) are low, and fine-needle aspiration of cold and indeterminate nodules is recommended to avoid unnecessary surgery. Although a dominant cold or indeterminate nodule of a clinically apparent multinodular goiter is less likely to be malignant than a similar solitary lesion, the malignant potential of such dominant nodules should still be assessed by fine-needle aspiration. Moreover, the demonstration on scan of multiple cold areas that do not correspond to palpable nodules does not lower the risk of malignancy in a patient with a clinically solitary nodule.[9]

Two other imaging modalities are commonly used to evaluate solitary or dominant thyroid nodules. X ray fluorescent scanning quantitates the iodine content of the nodule and adjacent extranodular thyroid tissue. The iodine content of the two regions is compared and an iodine content ratio of less than 0.6 has an equal sensitivity (97%) and better specificity (60%) than radionuclide thyroid imaging in the detection of thyroid cancer.[10] Fine-needle aspiration could then be performed on all nodules with iodine content ratios of less than 0.6 to determine which nodules require operation. However, because of their limited applicability only two fluorescent scanning systems are currently functioning and they are no longer commercially available. High-resolution real-time small-parts sonography visualizes thyroid nodules as small as 1 mm as solid, mixed, or cystic lesions. True simple cysts of the thyroid are very rare and most cystic lesions contain some solid tissue as a result of cystic degeneration of a follicular adenoma. Because both benign and malignant lesions can have cystic components, distinction of a cyst from a solid mass is not clinically useful.[11] All lesions require fine-needle aspiration to better define a nodule's malignant potential. Because of its exquisite resolution, real-time ultrasonography detects micronodules in 40% to 50% of patients with clinically apparent solitary nodules. However, as discussed above, the presence of these micronodules does not lower the risk of malignant disease as does the presence of a clinically apparent multinodular goiter. Neither CT nor NMR are useful in differentiating benign from malignant nodules. Because of the low specificity of radionuclide thyroid imaging and real-time ultrasonography, some thyroidologists believe that fine-needle aspiration should be performed without prior imaging studies. Patients with positive or suspicious follicular cytology would then undergo radionuclide imaging to exclude the 40% of hyperfunctioning nodules that exhibit such cytologic findings. A study comparing these two approaches, fine-needle aspiration before or after radionuclide imaging in the same population, is currently in progress, but has not yet been reported.

Upper Mediastinal Masses

If chest X ray or CT imaging demonstrates an upper mediastinal mass, it most likely represents a substernal goiter, but metastatic lung cancer, thymoma, or other mass lesion need to be considered. Avoidance of thoracotomy or mediastinoscopy is possible by documenting functioning thyroid tissue within the mass lesion. ^{123}I or ^{131}I in the older patient are preferred over Tc-99m imaging because of the latter's persistence in the blood pool of the great vessels at 20 minutes post-tracer administration. Pinhole scintigraphy can induce a parallax error that causes deeply located thyroid tissue caudal to the suprasternal notch to appear cephalad to the notch on thyroid imaging. This error must be avoided by placing the pinhole perpendicular to the suprasternal notch instead of over the thyroid bed and placing the patient's head in the same position as that used for chest radiography.[12] Overall accuracy should exceed 95% if such details are observed.[13] Most intrathoracic goiters exhibit anatomic continuity with a cervical portion of the gland. CT studies of confirmed mediastinal goiter reveal five characteristic features that strongly suggest the diagnosis of mediastinal goiter. Most intrathoracic goiters exhibit anatomic continuity with the cervical thyroid, focal calcifications, relatively high CT numbers, a rise in CT number after bolus administration of iodinated contrast material, and prolonged enhancement after contrast administration.[14] In patients with compressive symptomatology, CT or NMR offer important information in regard to tracheal or esophageal displacement or compression not available from radionuclide imaging.

Plummer's Disease Versus Graves' Disease

Approximately 16% to 20% of patients with hyperthyroidism referred for ^{131}I therapy demonstrate a nodular goiter to palpation. Prior to ^{131}I therapy, pinhole thyroid scintigraphy with ^{123}I or Tc-99m is useful in the differentiation of the various causes of hyperthyroidism.[15–16] In the absence of class II or III ophthalmopathy, Graves' disease superimposed on a previous multinodular goiter cannot be distinguished from Plummer's disease without scanning (Fig. 7.1). In patients with Graves' disease and a solitary or dominant cold nodule, fine-needle aspiration of the nodule to determine its cause can be performed prior to ^{131}I therapy or after therapy if the nodule persists. Plummer's disease may manifest itself as a single, autonomously functioning thyroid adenoma or multiple hyperfunctioning areas on scan. If needed, repeat thyroid imaging after exogenous thyrotropin stimulation will show a change in scan appearance in Plummer's disease but not in Graves' disease. CT ultrasound and NMR are not useful in this differentiation. The value of such selective use of scanning in the hyperthyroid population will obviously be limited in an endemic goiter region.

Thyroid Remnant Postthyroidectomy

Recurrence of nodular goiter or enlargement of residual thyroid tissue following thyroidectomy for benign disease is readily evaluated by pinhole thyroid scintigraphy. Such recurrences are seen more commonly if the

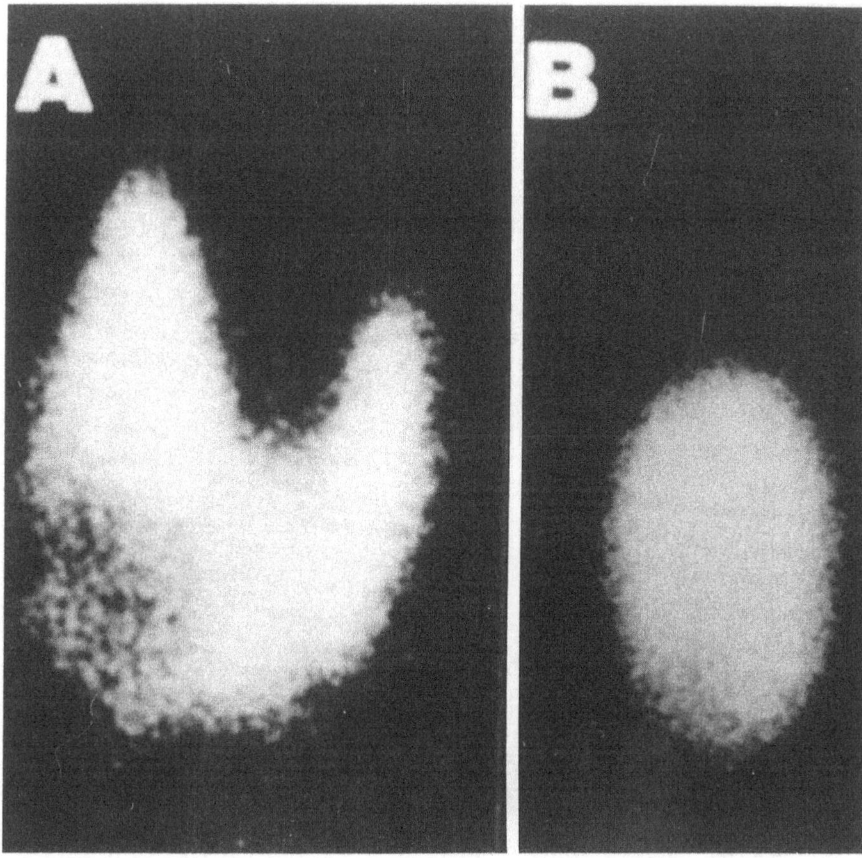

FIGURE 7.1. Hyperthyroidism in patients with nodular goiter. A: 3-cm right lower pole cold nodule in patient with Graves' disease. Fine-needle aspiration prior to ^{131} therapy confirmed colloid nodule. B: 3-cm right-lobe hot nodule in patients with Plummer's disease.

patient has not been taking thyroxine replacement after surgery. If the palpable thyroid tissue is functioning by scan consistent with compensatory hyperplasia, reduction in size can usually be achieved by institution of thyroxine replacement. If the tissue is nonfunctioning, fine-needle aspiration should be performed to clarify its nature. If the tissue recurrence is in a patient with Graves' disease or previous multinodular goiter, the tissue may be nonsuppressible (Fig. 7.2).

Residual or Metastatic Thyroid Cancer

^{131}I imaging has played a role in the follow-up and therapy of well-differentiated thyroid cancer for many years. In those patients felt to be at increased risk for recurrence or mortality from papillary or follicular

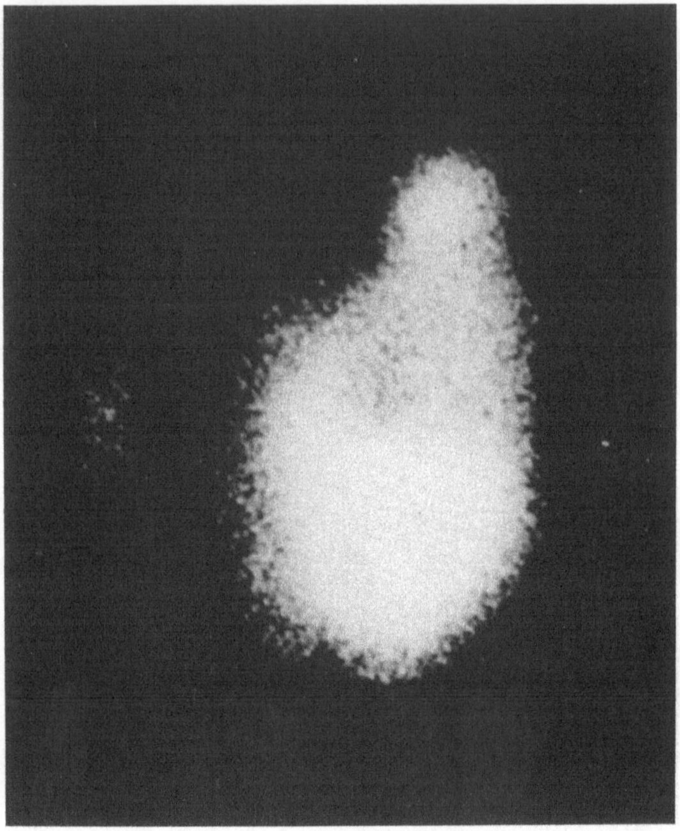

FIGURE 7.2. The patient noted gradual enlargement of left lobe of thyroid on thyroxine therapy following previous right lobectomy for colloid nodule. Scan (anterior view) demonstrates autonomously functioning thyroid adenoma.

thyroid cancer, it is customary to obtain ^{131}I scan (with 2 to 10 mCi dose) at 6 weeks postoperatively (Table 7.1).

^{131}I thyroid bed ablation or treatment of local or metastatic disease can then be performed if sufficient ^{131}I uptake is present. Following the administration of an ablation or treatment dose of ^{131}I, a posttherapy scan should be obtained at 3 to 7 days to better define the extent of disease. Such posttherapy scans demonstrate new lesions in as many as 16% of the patients who have undergone standard 2 to 10 mCi diagnostic studies.[17] Posttherapy scans also offer prognostic information since patients without metastatic disease visualized on scan are highly unlikely to require subsequent ^{131}I therapy in the future as compared with those with visualized metastatic disease. It is customary in many centers to follow thyroidectomized ^{131}I-treated patients with repeat ^{131}I whole body scans at 1- to 5-

TABLE 7.1. Patients at increased risk for recurrence or mortality.

1. Age less than 20 with papillary or follicular cancer
2. Age 20 to 40 with papillary or follicular cancer
a. Known residual tumor postoperatively
b. Regional or distant metastases
c. Recurrent disease
3. Age greater than 40
a. Papillary cancer >1 cm
b. All follicular cancer

year intervals. Such patients need to be withdrawn from their thyroid hormone for 6 weeks before readministration of diagnostic doses of ^{131}I. There is now considerable evidence that serum thyroglobulin is useful in monitoring these patients.[18-20] With a sensitive immunoassay, recurrence of thyroid cancer can be detected by monitoring serum thyroglobulin levels while the patient continues to take his or her thyroxine hormone. If the serum thyroglobulin value is elevated, thyroid hormone can be withdrawn and the patient scanned to detect recurrent disease. If serum thyroglobulin values remain suppressed, ^{131}I body scans may not need to be performed at 1- to 5-year intervals. Preliminary information suggests that the majority of patients with an elevated serum thyroglobulin level but negative routine ^{131}I scans (2- to 10-mCi dose) will show abnormal uptake following treatment doses (100 to 150 mCi). Additional patient studies need to be performed to clarify this issue.

Neonatal Hypothyroidism

Persistent neonatal hypothyroidism occurs with an incidence of 1 per 2,700 live births as detected by centralized laboratory screening programs. Thyroid scanning with ^{123}I or Tc-99m at the time of detection differentiates those infants who are athyreotic, dyshormonogenic, or ectopically located.[21] Those with dyshormonogenesis (usually goitrous) should have genetic counseling provided to the parents. Those infants who appear athyreotic should have their thyroid hormone withdrawn at age 2 to 3 years to confirm persistent disease, while those with dyshormonogenesis or ectopic thyroid need not undergo thyroid hormone withdrawal in the future (Fig. 7.3). ^{123}I is preferred over Tc-99m for better image quality in these infants, but the latter is acceptable if ^{123}I is not readily available.

Midline Neck Masses

Thyroid imaging with ^{123}I or Tc-99m is useful in excluding ectopic thyroid tissue in patients with midline neck masses. The vast majority of such lesions represent thyroglossal duct cysts that contain no functioning thyroid tissue. If a normal thyroid gland can be palpated in the proper location, imaging of a midline neck mass can be obviated.

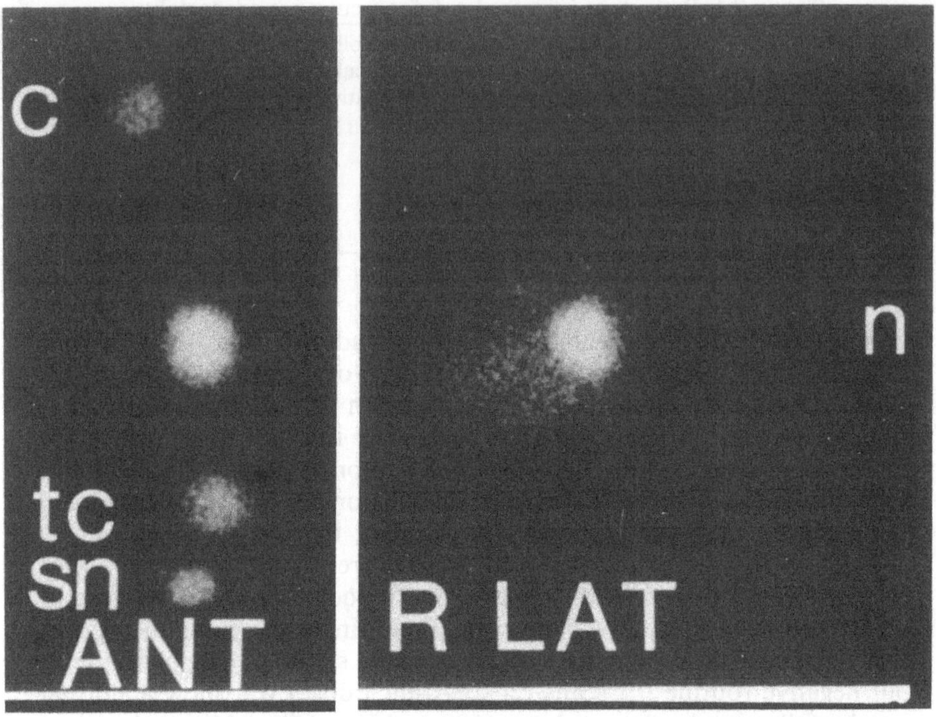

FIGURE 7.3. Left: Anterior view shows Tc-99m uptake in lingual thyroid of a 20-day-old infant with mild TSH elevation. Right: Right lateral view shows Tc-99m uptake at base of tongue. tc, thyroid cartilage; sn, sternal notch; c, chin; n, nose.

Summary

The volume of thyroid imaging has gradually declined over the past decade with the advent of fine-needle aspiration, sensitive serum thyroglobulin assays, and a better understanding by both imaging and referring physicians of the role of thyroid scanning in thyroid disease evaluation. Further decline in thyroid imaging may well occur as its utility becomes better defined. If fine-needle aspiration of solitary or dominant nodules becomes established as the preferred initial step in evaluation, the volume of thyroid imaging for this indication will fall as the appropriate expertise in obtaining the specimen and cytologic interpretation becomes more widely available. If serum thyroglobulin values obtained while on thyroxine therapy are shown to be as or more sensitive than ^{131}I scanning in the detection of thyroid cancer recurrence, ^{131}I whole body scanning will be required less often in the follow-up of thyroid cancer patients. In order to provide optimal patient care and service to the referring physician, the imaging physician needs to keep abreast of developments in other fields that enable better evaluation of patients with thyroid disease.

Discussion by Dr. Joel I. Hamburger

Dr. Freitas describes the traditional use of bovine thyroid-stimulating hormone (TSH) to stimulate or reactivate function in the suppressed extranodular tissue of a thyroid gland containing a hot autonomously functioning nodule. The use of TSH is no longer practical because the cost has escalated to $175 for a 10-unit ampoule. Fortunately, equivalent information can be obtained with a thallium image at less cost and without the potential risks of the injection of a foreign protein.

Dr. Freitas advised imaging 7 to 14 days after ^{131}I therapy for thyroid cancer because extra foci of tumor may be identified (beyond those "seen" with the 2- to 10-mCi tracer dose), and because the absence of additional foci makes it unlikely that further ^{131}I therapy will be needed. I have not used this technique because I do not think the findings are of any value. Additional tumor foci that had not been "seen" with tracer doses of 2 to 10 mCi do not take up enough ^{131}I to make that treatment useful. Furthermore, whenever I have had to re-treat patients with ^{131}I it has been because foci of uptake easily seen with the initial tracer doses were not ablated by the first therapeutic dose of ^{131}I. I wonder if Dr. Freitas can recall even one case that had additional foci of ^{131}I uptake seen after the therapy dose that were not seen on the initial tracer dose, who then had to be (or even could be) re-treated at a later date for uptake in those foci seen only after the therapy dose?

What Dr. Freitas has recommended is widely used, and if it provides no useful information, as I suspect, perhaps it is time to stop doing it in the interest of economy.

Response by Dr. John E. Freitas

Bovine TSH is expensive and rarely needed in the differentiation of Graves' disease (superimposed on colloid nodular goiter) from toxic multinodular goiter (Plummer's disease). However, it should provide the answer, whereas I do not believe that thallium will clarify this setting. I have not needed to use bovine TSH to differentiate toxic Autonomously Functioning Thyroid Adenoma (AFTA) from Graves' disease, but I agree with Dr. Hamburger that a thallium scan should answer that question.

The recent literature does not support the statement "Additional tumor foci that had not been 'seen' with tracer doses of 2 to 10 mCi do not take up enough ^{131}I to make that treatment useful." The work of Schlumberger, Pacini, and others has shown that lung lesions seen only on scans obtained with 100 mCi of ^{131}I can be eradicated as confirmed by normalization of subsequent scans and serum thyroglobulins. We have had only two cases (since initiating posttherapy scanning in 1982) in which new lesions were seen in scans that received subsequent treatment. I feel that posttherapy scans afford useful information on individual patients in terms of initial staging and identification of high-risk patients.

Discussion by Dr. James C. Sisson

With regard to the approach to the solitary or dominant thyroid nodule, I agree that many cysts contain a substantial solid component but not necessarily most. Some nodules may be purely cystic, a consequence of a hemorrhage, or contain a very small solid component. Ultrasound will help to distinguish the pure cysts from cysts that have a nodule of 1 cm or more, particularly when: 1) aspiration of the cyst is incomplete and residual fluid impairs palpation, and 2) the cyst fluid exhibits only debris and old blood (as is usually the case).

Differentiation of Plummer's disease (autonomous multinodular toxic goiter) from nodular Graves' disease is not done with precision, but, more importantly, is usually unnecessary. To give exogenous thyrotropin to patients with hyperthyroidism as suggested is to incur a double risk: 1) reaction to the foreign protein of bovine TSH, and 2) aggravation of hyperthyroidism. Imaging is appropriate to define multiple nodules in a toxic gland. Experience suggests that the nodules make the toxic gland, whether Plummer's or Graves' disease, more resistant to radioidine therapy. This is the only knowledge desired.

Biopsy of a dominant hypofunctioning nodule in a Graves' disease gland is suggested while hyperthyroidism persists. Aspiration biopsies, particularly those of many needle passes, cause hematomas. The hematomas induced by passing needles into the highly vascular gland of Graves' disease may be larger, and such hematomas, through compression, could aggravate hyperthyroidism to dangerous levels. Why not wait until euthyroidism has been restored before the biopsy?

You have said that a patient less than 20 years of age with papillary or follicular carcinoma is at greater risk than a patient with a comparable stage of the cancer who is 20 to 40 years of age. Is there any evidence to support this categorization?

It is not clear why imaging is performed in patients with midline neck masses since "the vast majority of such lesions represent thyroglossal duct cysts that contain no functioning tissue."

Response by Dr. John E. Freitas

There are multiple approaches that can be selected in treating the Graves' disease patient with a coincidental cold nodule. Fine-needle aspiration (FNA) prior to ^{131}I therapy has been safe in our hands and allows the patient to be treated definitively within a few days with ^{131}I if the cytology is benign. If there was real increased risk of complications from FNA in the hyperthyroid gland, rendering the patient euthyroid first would be appropriate, but we have not found this necessary to date.

Patients less than 20 years of age exhibit a high prevalence of regional or distant metastases at initial presentation.[22,23] The presence of regional metastases increases the local recurrence rate as compared to older patients, but is very responsive to surgical resection and ^{131}I. Lung metastases are present in up to 20% of patients (age <20 years) and are readily eradicated at this micronodular stage when chest X rays are often negative.

As stated, if a normal thyroid gland is palpated, scanning can be omitted. Scanning is useful in demonstrating that the midline mass represents the patient's ectopic thyroid gland. These masses usually shrink with thyroid suppression.

Discussion by Dr. Ian D. Hay

In patients at increased risk of thyroid cancer recurrence, we have since 1985 successfully used high-resolution (10 MHz) real-time sonography to identify recurrent tumor in the thyroid bed or adjacent cervical lymph nodes. These recurrent lesions have usually been found in patients with increased serum levels of Tg or calcitonin[24] and often in circumstances where an ^{131}I scan with a 3-mCi dose had been read as negative. Sonography can detect impalpable disease that can be biopsied under ultrasound guidance. A negative pathologic report in this circumstance can avoid an unnecessary cervical reexploration.

What is Dr. Freitas's opinion regarding the role of whole body thallium-201 scanning in the follow-up of patients with thyroid carcinoma? Has Dr. Freitas had any consistent success with newer imaging agents in visualizing metastases from medullary thyroid cancer?

Response by Dr. John E. Freitas

I have had little experience with the use of high-resolution real-time sonography in the follow-up of thyroid cancer patients. I find Dr. Hay's comments quite interesting.

^{201}Tl scanning in the follow-up of patients with differentiated thyroid cancer has been disappointing to us. It has not proven to be a sensitive method of detecting recurrent or metastatic disease in our patients. However, we have noted thallium uptake in some regional metastases that have been serum thyroglobulin positive and ^{131}I negative.

I do not have access to any of the newer imaging agents for medullary thyroid cancer or sufficient patients in whom to test them.

Discussion by Dr. Leonard Wartofsky

In our experience, many large substernal goiters are nonfunctional based upon radioiodine trapping. In such cases, have you still seen a rise in CT number after administration of iodinated contrast?

You obtain a scan image after therapeutic doses, how do you follow up lesions seen on the posttreatment (large dose, 150 mCi) scan that were not visualized on the routine (2 to 10 mCi) neck and chest survey since one would not expect to see these lesions on a subsequent (2 to 10 mCi) survey either?

I am confused about the utility of scanning for midline masses. Couldn't a midline thyroglossal duct cyst be present whether or not a normally palpable thyroid gland was present, and how does the presence or absence of uptake in the thyroglossal duct tissue help in ruling in or out an associated malignancy?

Response by Dr. John Freitas

We have not had any substernal goiters that did not trap ^{131}I. I do not know if such goiters that do not trap radioiodine would exhibit a rise in CT number.

We routinely obtain a posttherapy scan at 3 to 7 days post ^{131}I administration. The majority of lesions seen only on posttherapy scans are not visualized on subsequent posttherapy scans (after second or third ^{131}I therapies). In other patients whose subsequent 5- to 10-mCi scans are negative, we rely on a sensitive serum thyroglobulin assay to decide our next step. If the serum thyroglobulin value is normal, no further scanning is performed. If the serum thyroglobulin value is elevated, we would probably administer a 150-mCi ^{131}I dose and obtain a posttherapy scan (we have not had such a patient yet).

Our reason for scanning patients with midline masses is to detect the occasional patient whose mass represents the only functioning thyroid tissue. Such patients could be treated with thyroid hormone with significant regression of the mass and no operation would be necessary. If thyroid tissue is palpable in the normal location, the midline mass has always been nonfunctional in our experience and scanning is not useful.

Discussion by Dr. John T. Dunn

I commend the author for his restrained plea for using thyroid imaging techniques selectively and intelligently rather than excessively.

In neonatal hypothyroidism, I would raise these questions:

Is it worthwhile to scan, since all will be treated with T_4?

Is it really important to distinguish between athyreosis and dyshormonogenesis? You would be alert for hypothyroidism in subsequent pregnancies in either event, and dyshormonogenesis per se would not be a usual reason to avoid later pregnancies.

Isn't it easier to distinguish between these causes by the serum Tg than by the scan?

Response by Dr. John E. Freitas

If one does not scan infants prior to initiating T_4 treatment, all patients would need to have their therapy stopped at 2 to 3 years of age to determine which ones have permanent versus transient hypothyroidism. After scanning, only those with apparent athyreosis would need to be restudied at age 2 since some of these will have normal thyroid function on repeat testing.

The serum thyroglobulin value may not distinguish athyreosis from dyshormonogenesis reliably since the latter may also exhibit defective thyroglobulin synthesis. I am not aware of any publication that has shown good distinction between these groups by serum thyroglobulin assays.

Discussion by Dr. John T. Dunn

I put the question of posttherapy scans to Drs. David Becker and Harry Maxon, both experts on radioiodine therapy. They were of the opinion that if the uptake were so small that it could only be seen on a scan after 150 mCi of ^{131}I, it is likely that we wouldn't be effective in treating such lesions.

Discussion by Dr. James C. Sisson

I am not sure that further radioiodine treatment would be worthwhile for uptake that was only seen on an image made with 100 mCi or more. On the other hand, it does put the patient's stage of disease into a different context, especially if we were to see uptake in the spine, the lung, or some other remote area. I think this would call for more aggressive follow-up in terms of subsequent scanning.

Discussion by Dr. Paul Lo Gerfo

Despite all the data published about routine postoperative scanning, is there any evidence that the patients who are scanned and treated do any better than those who are not scanned and not treated (i.e., only treated if Metastases (mets) become evident).

Has anyone followed patients who have undergone routine body scans to see if ^{131}I has done more harm than good?

I am curious as to how many of the panel use thyroid scans (images) in the evaluation of thyroid nodule patients. I see many thyroid nodules, and some have been scanned before I evaluate them; for the others, as many as 200 per year, I may not order scans more than once in 5 years. I am not sure that scans play an important role in the workup of this disease. Am I the only one with this opinion here?

Response by Dr. John E. Freitas

The value of radioiodine in the treatment of patients with well-differentiated thyroid cancer is best demonstrated when one looks at those patients with metastases outside the neck.[25] These patients exhibit a marked decrease in survival as a group as compared to an age-matched control population. Treatment of these patients with radioiodine prolongs survival twofold and enables cure in the majority of patients with mediastinal or lung metastases. Several studies have addressed the importance of early detection of lung metastases by whole body radioiodine scanning as opposed to waiting until chest X rays become positive. Complete remissions of lung metastases were achievable in 64% of the patients with a normal chest X ray yet positive ^{131}I scans as compared to only 8% of patients with positive chest X ray and radioiodine scans.[26] Proper management of thyroid cancer patients requires early detection of metastatic disease so that such metastases can be treated as early as possible at the time the volume of tumor is at a curable stage.

Radioiodine has proven benefit for some patients with differentiated thyroid cancer, as discussed above. In many large series of patients with differentiated thyroid cancer treated with radioiodine and followed by routine whole body scans, there is no increase in leukemia or second cancers, as compared to a matched control population. Such results contrast sharply with the 90- to 170-fold increase risk of acute leukemia at 2 years after the treatment of carcinoma of the ovary with chemotherapy or the 20-fold increased risk of developing a second cancer at 4 years after the treatment of Hodgkin disease with X ray therapy and cancer chemotherapy.[27,28] Follow-up radioiodine scanning with 2 to 3 mCi of ^{131}I delivers a whole body radiation dose of 0.7 to 1 rad per study. This is a lower whole body radiation exposure than many other commonly performed radiology procedures. There is no evidence to support any deleterious effect from these low doses of radioiodine.

Discussion by Dr. Leslie J. DeGroot

I guess I order scans more often than every 5 years, but certainly have relied upon fine-needle aspiration cytology as the main-line approach to the single thyroid nodule, without scans. If the cytology is suspicious, I send the patient to surgery without a scan, so I agree with Dr. Lo Gerfo on that point.

Also, I haven't found any use for scanning people for whom I suspect Graves' disease. If I am concerned about a multinodular goiter I do get a scan, although I don't think it actually changes our therapy with ^{131}I much.

With regard to scanning thyroid cancer patients after ^{131}I therapy, we tried it for several years, but found the results very confusing and gave up doing them. For example, after the large therapy dose we often see small foci of uptake that were not seen subsequently with conventional scan doses of 2 or 5 mCi. We decided the information from the post-therapy scans wasn't useful so we quit doing it.

References

1. Hamburger JI, Miller JM, Garcia M, et al: Is thyroid imaging an overutilized diagnostic procedure? In: Hamburger JI, Miller JM, eds. Controversies in Clinical Thyroidology. New York: Springer-Verlag, 1981. pp 1–20.
2. Reasner CA, Isley WL: Thyroid scanning in patients with suspected thyroid disease. Ann Intern Med 1985;102:209.
3. Van Herle AJ: The thyroid nodule. Ann Intern Med 1982;96:221.
4. Sandler MP, Patton JA: Multimodality imaging of the thyroid and parathyroid glands. J Nucl Med 1987;28:122.
5. Silverberg E: Cancer statistics. CA 1987;37(1):2.
6. Vander JB, Gaston EA, Dawber TR: Significance of solitary nontoxic thyroid nodules. N Engl J Med 1954;251:970.
7. Sokol JE: The problem of malignancy in nodular goiter: Recapitulation and challenge. JAMA 1959;170:405.
8. Okerlund MD, Sommers J, Grell T, et al: Clinical importance of the "indeterminate" nodule in thyroid imaging: The commonest cause of the false negative thyroid imaging study. J Nucl Med 1980;21:31.
9. Taylor S: Sporadic nontoxic goiter. In: Werner SC, Ingbar SH, eds. The Thyroid, 4th ed. New York: Harper & Row, 1978, 505–512.
10. Patton JA, Sandler MT, Partain CL: Prediction of benignancy of the solitary "cold" thyroid nodule by fluorescent scanning. J Nucl Med 1985;26:461.
11. Butch RJ, Simeone JF, Mueller TR: Thyroid and parathyroid ultrasonography. Radiol Clin North Am 1985;23(1):57.
12. McKitrick WL, Park HM, Kosegi JE: Parallax error in pinhold thyroid scintigraphy: A critical consideration in the evaluation of substernal goiters. J Nucl Med 1985;26:418.
13. Park, Carver RD, Siddiqui AR, et al: Efficacy of thyroid scintigraphy in the diagnosis of intrathoracic goiter. Am J Roentgenol 1987;148:527.
14. Glazer GM, Axel L, Moss AA: CT diagnosis of mediastinal thyroid. Am J Roentgenol 1982;138:495.
15. Ripley SD, Freitas JE, Nagle CE: Is thyroid scintigraphy necessary before ^{131}I therapy for hyperthyroidism? Concise communication. J Nucl Med 1984;25:664.
16. Fogelman I, Cooke SG, Maisey MN: The role of thyroid scanning in hyperthyroidism. Eur J Nucl Med 1986;22:397.
17. Bestagno MF, Paglania R, Mavra G: The use of radioiodine in the management of the differentiated thyroid cancer. J Radioanalyt Chem 1981;65:239.
18. Coakley AJ, Page CJ, Croft B: Scanning dose and detection of thyroid metastases. J Nucl Med 1980;21:803.
19. Schlumberger M, Tubiana M, DeVathaire F, et al: Long-term results of treatment of 283 patients with lung and bone metastases from differentiated thyroid carcinoma. J Clin Endocrinol Metab 1986;63:960.

20. Pacini F, Lippi F, Formica N, et al: Therapeutic doses of iodine-131 reveal undiagnosed metastases in thyroid cancer patients with detectable serum thyroglobulin levels. J Nucl Med 1987;288:1888.
21. Klein RS: Infantile hypothyroidism: Then and now. Results of neonatal screening. Pediatrics 1985;15:1.
22. Goepfert H, Diebtel WJ, Samaan NA: Thyroid cancer in children and teenagers. Arch Otolaryngol 1984;110:72.
23. Leeper RD: Thyroid cancer. Med Clin North Am 1985;69:1079.
24. Gorman B, Charboneau JW, James EM, et al: Medullary thyroid carcinoma: Role of high-resolution US. Radiology 1987;162:147.
25. Beierwaltes WH, Nishiyama RH, Thompson NW, et al: Survival time and "cure" in papillary and follicular thyroid carcinoma with distant metastases: Statistics following University of Michigan therapy. J Nucl Med 1982;23:561.
26. Schlumberger M, Tubiana M, De Vatharie F, et al: Long-term results of treatment of 283 patients with lung and bone metastases from differentiated thyroid carcinoma. J Clin Endocrinol Metab 1986;63:960.
27. Beierwaltes WH: New horizons for therapeutic nuclear medicine in 1981. J Nucl Med 1981;22:549.
28. Chabner VA: Second neoplasm—A complication of cancer chemotherapy. N Engl J Med 1977;297:213–214.

8

Postoperative Evaluation of Thyroid Cancer Patients for Adequacy of Surgery and for Need of ^{131}I Therapy

JAMES C. SISSON

The idea that surgical excision is the first line of treatment of thyroid cancers, and particularly of well-differentiated thyroid cancers, provokes little controversy. However, most physicians will state that thyroidectomy should not be the sole therapy considered for the patient with thyroid cancer. Yet, the type of operation performed greatly affects what can be done further and, therefore, the surgeon should be a part of a team of oncologic physicians. Surgeons should not undertake treatment of thyroid cancer alone.

This communication will deal only with the well-differentiated thyroid cancers: papillary (or papillary-follicular) and follicular carcinomas. Most of these neoplasms concentrate ^{131}I to some degree, and the use of this function for diagnosis and treatment is a major focus of this discussion.

The Study of Patients with Thyroid Cancer by ^{131}I Scintigraphy

Concepts

There is general agreement on the concepts governing the postoperative evaluations of patients with well-differentiated thyroid cancers. We should ascertain:

1. If there is a threat to health and life of the patient by persisting neoplasm;
2. If ^{131}I in therapeutic doses would reduce any determined threat;
3. If ^{131}I in therapeutic doses would pose a lesser threat than the cancer to the patient.

Whereas these concepts are easily accepted, application of them to an individual patient is often difficult for want of specific guidelines. Following are data that will serve to direct the application of the concepts.

Threat of Thyroid Cancer

A number of types of surgical operations are used for thyroid cancer; vocal advocates claim success for each approach.[1-3] The threat of thyroid carcinoma is sometimes obscured by the controversy over which operation should be performed. Unfortunately, there is no randomized comparison to guide us at this point. Nevertheless, factors clearly affecting prognosis have been identified: histology of the neoplasm, extent of tumor at time of diagnosis, and age at diagnosis. How one proposes to deal with these individual risk factors will determine the type of surgical excision that should be undertaken. Thus, we must examine each of these factors that threaten the health and life of patients with thyroid cancer.

Rates of death from papillary thyroid cancer vary in reports but rarely exceed 10%[1-4] (Table 8.1). On the other hand, follicular thyroid cancer appears to be a substantial risk to life[3-5] (Table 8.1); it is the angioinvasive follicular cancers that cause a greater death rate than papillary and non-invasive follicular tumors. In addition, the Hurthle cell tumor, a variant of the follicular cancer, and the "tall cell" papillary carcinoma are more lethal forms of the well-differentiated thyroid cancers; but these types of neoplasm usually concentrate little or no radioiodine and will not be considered in this chapter.

When prognosis is related to extent of disease, the standard TNM classification does not serve well for thyroid cancers. For papillary carcinomas, the size of the primary tumor correlates with future events: a lesion (or lesions) of less than 1 cm in diameter rarely causes death (only one of 182 patients died from cancer initially observed at this stage) or distant metastases, but morbid events become more probable when the primary tumor is greater than 3 cm.[2] However, regional lymph node

TABLE 8.1. Death from well-differentiated thyroid cancer.

Reference	Years of entry	Years of follow-up (median)	Number of Patients	Death (%)
Papillary				
4 (U.S. Air Force)	1962–72	10	576	1.0*
1 (Mayo Clinic)	1946–70	18	859	6.5†
2 ‡(M.D. Anderson)	1951–75	(until 1981)	574	9.1*
3 (Lahey Clinic)	1931–70	18	761	10.1†
Follicular				
5 (U.S. Air Force)	1962–72	9	214	1.9*
2 (M.D. Anderson)	1951–75	(until 1981)	108	15.7*
3 (Lahey Clinic)	1931–70	18	220	17.7†
6 (University Michigan)	1962–82		37	16.2*

*^{131}I given as treatment to many patients.
†Little or no ^{131}I given as treatment to patients in these series.
‡Combining papillary and mixed thyroid cancer.

metastases do not connote a poorer prognosis.[1,3] In fact when papillary carcinoma occurs in a patient less than 30 years of age, it is frequently found in lymph nodes; yet such a patient often exhibits the best outcome in terms of health and survival.[1,3]

Nonetheless, the presence of either local invasion or distant metastases is an indicator of poor outcome. Papillary cancers reach distant sites by lymphogenous routes while the follicular carcinomas disseminate hematogenously. For either histologic type of neoplasm, these remote metastases are frequently lethal,[1,2] and deposits in bone and macronodular disease in lung are more dire than micronodular metastases in lung.[7]

When the diagnosis of thyroid cancer is made after the age of 40 to 50 years, the prognosis is less favorable than when the neoplasm arises at an early stage of life.[1–4] The reasons for the effects of age are not known, but the neoplasms of the older patients tend to appear less well-differentiated and concentrate less radioiodine than those in younger subjects.

One group found that men tolerated thyroid cancer less well than women,[1] but other investigators observed the opposite.[3] If a difference in prognosis between the sexes exists, it is small and does not bear on decisions of management.

From the above data, well-differentiated thyroid cancers can be classified according to risk as shown in Table 8.2.

Efficacy of ^{131}I Treatment

A number of articles extol the value of ^{131}I treatments for thyroid cancer.[2,4–6,8,9] Unfortunately, there have been no rigorously controlled studies to determine the efficacy of this form of treatment. Certainly, in sufficient concentration, ^{131}I will destroy thyroid tissue, and most physicians agree that radioiodine treatments have benefited some patients. Nevertheless, there has been no consensus on the indications for ^{131}I therapy, including what roles should be played by the different stages of the thyroid cancer and the dosimetry of ^{131}I in the choice of this treatment.

TABLE 8.2. Risks from well-differentiated thyroid cancers.

	Histologic type	
Relative risk*	Papillary	Follicular
Little or none	Largest primary <1 cm	
Low†	Primary 1–3 cm, with or without lymph node metastases	Primary up to 3 cm, and no angioinvasion
Moderate-high	Primary >3 cm, or local invasion, or distant metastases	Primary >3 cm, or angioinvasion or local invasion by any primary tumor, or distant metastases

*Although age at time of diagnosis is clearly an independent factor in risk, the overall threat of well-differentiated thyroid cancer is dominated by histologic type and extent of disease.
†Death rate from thyroid cancer will be less than 5%.

After surgical treatment, patients should at least be considered candidates for radioiodine treatment. A patient could be at an extreme in the spectrum of disease (e.g., one manifesting distant metastases (in lung and/or bone) that appear to be avid for ^{131}I) and, for this case, there can be little disagreement with the indication for therapeutic radioiodine. This type of patient requires help, and, although of unproven efficacy, radioiodine appears to be the only reasonable treatment. However, for other patients with thyroid cancer, decisions will be more controversial and will reflect the bias of the clinician.

Relative Safety of ^{131}I

Many patients have received ^{131}I in substantial quantities; 100- to 300-mCi doses have been given, sometimes on multiple occasions.[2,4,5,6,8,9] In general, radioiodine has been safe.[10,11] At very high doses, bone marrow suppression has occurred, and a patient with lung metastases developed pulmonary fibrosis subsequent to radioiodine treatment.[12] Since the salivary glands concentrate ^{131}I, dysfunction in these organs occurs in some patients,[13] but the actual prevalence of xerostomia following radioiodine is unknown.

A major concern of ^{131}I therapy is for leukemia.[14] Radiation is known to induce leukemia,[15] and anecdotal cases of leukemia arising after ^{131}I treatments, some following large cumulative doses, have been reported.[16] No cause-and-effect relationship has been clearly established for ^{131}I and leukemia, and the risk, if any, must be of low magnitude. Still, the use of radioiodine should take into account the possibility of inducing a hematopoietic malignancy.

What Are the Indications for Studying Patients with Thyroid Cancer; Who Should Be Investigated?

Scintigraphic studies should provide data that will be decisive in whether to give or not give ^{131}I as therapy. If, at the outset, a patient is not deemed a candidate for radioiodine treatment, then scintigraphy ought not be performed. A younger age at the time of diagnosis of well-differentiated thyroid cancer should not alone be a reason for omitting the studies. Although the vast majority of patients in whom the cancer was detected before 40 years of age will live a normal span, some will have recurrences of tumor that alarm the affected individual,[1–4] and a few, especially those in the more advanced risk categories, will die of the neoplasm.[1–4,6] Factors that bear importantly on the decision to investigate are: histologic type of neoplasm, extent of disease, extent of the thyroidectomy, and suspicion from clinical and laboratory evidence of residual or recurrent cancer.

Histologic Type of Cancer

When the histology of the cancer is papillary or follicular, ^{131}I scintigraphy will be a reasonable choice regardless of the amount of follicle or colloid formation. Whereas, the concentration of ^{131}I is highest in the most well-differentiated (normal-appearing) tumors, some cancers with much less differentiation will still sequester sufficient ^{131}I to enable therapeutic radiation to be delivered. Moreover, the histologic pattern of the carcinomas may vary from site to site.[2,4,5,6,8,9] On the other hand, patients whose tumors are found to have undifferentiated components will virtually always have a rapidly progressive course that is not amendable to ^{131}I[17] and scintigraphy will be of no help to them.

Extent of Disease

When the primary tumor is papillary and is less than 1 cm in diameter, the cancer is called "minimal" and the prognosis for health and life is so good[1] that probably no treatment beyond excision is necessary. However, if lymph node metastases are associated with a "minimal" papillary tumor, which is frequently the case, then there is more concern about future effects from the cancer. My own approach would be to treat such small neoplasms as incidental (i.e., requiring neither study nor treatment other than thyroid hormone) if metastases were absent or only microscopic in size. In contrast, clinically apparent metastases are associated with recurrences[1] that alarm the patient, and, for individuals with minimal tumors and palpable nodes, I recommend resection of the nodes (but not a radical dissection) and a study for residual disease by ^{131}I scintigraphy. Probably patients who have follicular cancers that are less than 1 cm in diameter and express no angioinvasion could also safely forego further study and receive only thyroid hormone therapy.

All other stages of well-differentiated thyroid cancers deserve to be appraised by images made with ^{131}I. In fact, scintigraphic evidence is often essential in establishing the stage and extent of these carcinomas.

Extent of Thyroidectomy

Optimal images are obtained when there is an intense stimulation of function in the residual tissues. Such stimulation is best obtained by endogenous thyroid-stimulating hormone (TSH) that has risen to high levels consequent to the removal of nearly all of the normal thyroid gland. A scintigraphic search for metastatic deposits will be impaired by substantial amounts of normal thyroid tissue, particularly those associated with an uptake of 5% or more of the dose and a serum TSH of less than 30 $\mu U/ml$.[18] Before using ^{131}I to study a patient, thought should be given to removing any large residual of normal thyroid tissue by further surgical excision or by an ablative dose of ^{131}I (see below).

Suspicion of Residual or Recurrent Cancer

Following treatments, one may suspect residual and recurrent cancer in a variety of circumstances e.g., there may be clinically palpable tumor, X-ray abnormalities, or an unexplained concentration of serum thyroglobulin.[19,20] Even when no carcinoma appears to remain in a careful study, there is still a risk of recurrence that continues indefinitely.[1–4] For this reason, we have preferred to perform scintigraphy at about 5-year intervals indefinitely. These images would be repeated more frequently if other methods of evaluation raised the possibility of cancer. Whether annual measurements of thyroglobulin can satisfactorily replace periodic imagings as indices of recurrent and progressive thyroid cancer is unknown. It is possible that patients who have only a low risk from thyroid cancer (Table 8.2) could be followed safely by periodic reevaluation in which clinical examinations and assays of thyroglobulin are used as initial detectors of recurrent disease. However, the success of a treatment of cancer by radioiodine cannot be known except by a reevaluation employing all diagnostic means including scintigraphy, which may have given the only pretherapeutic evidence of tumor.

How Should Scintigraphic Studies Be Performed?

Optimal scintigraphic studies depend upon proper patient preparation and appropriate techniques of data acquisition.

Preparing the Patient

Most well-differentiated thyroid cancers will take up ^{131}I,[2,4] and probably most will exhibit an uptake that is responsive to stimulation by TSH. Thus, as noted above, the optimum uptake of ^{131}I by the neoplasms (optimum both for detection and treatment) depends upon an elevated level of circulating TSH, a state that is obtained by removing enough normal thyroid tissue to create rather profound hypothyroidism. Following a total or near-total thyroidectomy, clinically obvious hypothyroidism appears in 3 to 4 weeks, and probably the serum TSH is near maximal levels by 5 to 6 weeks.[21] Many physicians prefer to wait 6 weeks after thyroidectomy or withdrawal of thyroxine therapy, a time when TSH values are usually over 40 μU/ml and may be over 200 μU/ml in children, to perform scintigraphy. If triiodothyronine is substituted for thyroxine as treatment, withdrawal of this medication for 3 weeks appears to be sufficient;[18,24] patients will then still become symptomatic from hypothyroidism, but probably for a briefer period.

The TSH values may be unexpectedly low in a patient who: 1) did not discontinue thyroid hormone, at the proper time; 2) had taken unusually large maintenance doses of hormone (e.g., 0.3 mg or more per day of thyroxine); and 3) had substantial quantities of functioning thyroid tissue, either benign or malignant. In the first two instances, a more prolonged period of abstinence from hormone will give higher TSH results; in the last instance, large amounts of functioning tissue will be readily demonstrated by scintigraphy. Although injections of bovine TSH can stimulate human thyroid tissue, this approach incurs the possibilities inducing neutralizing antibodies to the hormone,[22] and a severe analphylactic reaction to the foreign protein in the occasional patient. Most physicians prefer to stimulate residual thyroid tissue via endogenous TSH.

The availability of iodides will also influence the uptake of tracer quantities of ^{131}I by thyroid cells. An adaptation occurs in normal and, presumably, neoplastic thyroid tissue to the circulating concentrations of iodide, and the lower the level of iodide, the greater the fraction of both stable and radioactive iodide sequestered.[23] Therefore, to attain the highest uptake of ^{131}I for a given dose, a low-iodine diet, usually less than 25 μg/d for a week, should be prescribed.[24,25] One report confirmed that a low-iodine diet increased the fractional uptake of ^{131}I by thyroid tissue, but also found a prolonged retention of the radioactivity in the body so that for a given dose of ^{131}I, both tumor and whole body received more radiation.[25] Nevertheless, the retained whole body radioactivity may well not be proportionally present in the tissues surrounding tumors and, in this event, the detection of neoplastic deposits by a given tracer dose will be enhanced by the low-iodine diet.

Techniques of Acquiring Data

Variety has characterized the tracer doses of ^{131}I for diagnostic scintigraphy. In recent years, doses of 1 to 10 mCi have been common. The larger the tracer dose the greater the likelihood of detecting small, elusive cancer deposits.[26] However, studies with mocks have shown that, using modern scintigraphic equipment, no reasonable tracer dose (even 30 mCi) can uncover all treatable metastases.[27] The magnitude of the tracer dose then becomes a compromise between desires for intense searches for thyroid carcinoma and lesser radiation delivered to the patient in diagnostic procedures that are likely to be repeated over the years.

We have decided that, for patients with low risk of cancer, 2-mCi tracer doses would suffice; this choice was especially appealing for a young patient who might require multiple scintigraphic studies in his or her lifetime. On the other hand, to patients with a moderate-high risk of cancer, we give 5- or 10-mCi tracer doses.

Following administration of the selected tracer dose, reasonable images may be created up to 3 days later, but those made at 2 or 3 days give a greater probability of detecting small tumors than images acquired at 24 hours.[28] The effective half-life of radioactivity in normal tissues is usually 1 day or less while that in the cancers, and particularly in the treatable cancers, is longer. Thus, the tumor-to-background ratio improves over time and, after only 2 or 3 days, the loss from the tumor would be unlikely to make it subliminal in an environment of low background radiation.

The screening of the patient should be by a modern gamma camera fitted with a high-energy parallel hole collimator and include images of the entire torso, the head, the femora, and the humeri. Because small but treatable tumors may be elusive, additional images, such as posterior views, should be added when areas have drawn increased suspicion from clinical or radiographic data. Also, acquisition of sufficient data through a pinhole collimator will better resolve any detected foci of radioactivity so that the anatomic relationships of the abnormality can be more accurately defined.

If one elects to give therapeutic doses of ^{131}I to tolerance, then formal dosimetry of normal tissues should also be carried out over 4 or more days.[9,29] Radioactivity retained in the blood and the entire body must be measured at intervals and the cumulative doses of radiation to each region calculated.

Also, if there is reason to believe that some of the neoplasm in the patient does not concentrate radioiodine, then images made with thallium-201 will add perspective to the appraisal of cancer.[30] When the image made with ^{201}TI indicates that only a part of a carcinoma sequesters ^{131}I, consideration should be given to therapies other than the radioiodine.

What Are the Meanings of Scintigraphic Results? What Actions Should Follow?

Two major factors govern the choice of dose of ^{131}I to treat patients who carry the diagnosis of thyroid cancer: the presence of what is almost certainly normal thyroid tissue, and the optimum dose to treat the identified cancer.

Doses of ^{131}I to Treat Normal Thyroid Tissue

Not uncommonly, a postoperative patient exhibits an easily measured amount of normal tissue (even an entire lobe) which the surgeon believes he or she cannot remove safely. When such a normal residual maintains the patient in a euthyroid or nearly euthyroid state, there will be little stimulation of cancer by TSH; hence, an optimum state for delivering ^{131}I to the neoplasm does not exist. Moreover, a major hazard of therapeutic doses of ^{131}I is the synthesis of ^{131}I-thyroxine by appreciable amounts of thyroid tissue; ^{131}I-thyroxine circulates with an effective half-life of 4 to 5 days and can irradiate the hemotopoietic system to dangerous levels.[12]

Some physicians prefer to give patients with large amounts of normal thyroid tissue standard treatment doses of ^{131}I, believing that the best opportunity to attack the cancer is with the first treatment. To them, the possibilities of subtherapeutic radiation to the cancer and damage of the bone marrow are acceptable. I prefer to give smaller doses, as low as 30 mCi, of ^{131}I to these patients. My concept is to reduce the normal thyroid tissue to hypofunctioning levels (and not necessarily cause its complete elimination), while at the same time imparting to the carcinoma levels of radiation that are too low to alter its functional mechanisms. A large, tumor-directed dose of ^{131}I would be given later when the neoplastic cells had been stimulated to optimal function by the consequentially elevated level of circulating TSH.

Small residuals of normal tissue pose a different problem. These are defined as small foci of radioactivity (overall uptake $<2\%$ of the tracer dose) observed on the postoperative scintigraphic image and are determined to represent benign thyroid tissue from the surgeon's description of the operation and the pathologist's interpretation of the excised specimens. For example, it is common practice for surgeons to avoid a region of normal gland near the recurrent laryngeal nerve, a site that surgeon and pathologist may agree is remote from any tumor within the resected tissue, but one that will be readily observed on the scintigraphic image. This definition of small foci of normal tissue differs from that used by others to justify ablation of any foci of tissue observed in the neck.[2,31] I see no reason to treat with ^{131}I small normal tissue remnants as I have defined them. Of course, ablation of such normal thyroid tissue has provoked controversy,[31-34] but no convincing case has been articulated for the position advocating treatment. If, in these instances, one aims to treat subliminal cancer with large doses of ^{131}I, then logic dictates giving all patients at least one treatment with radioiodine to destroy unseen tumors regardless of the presence of foci of normal tissue. Also, to administer, as others propose, 30 mCi to simply eliminate small residua of normal tissue, will neither properly attack subliminal cancer nor improve the ability to do so.

Selecting Doses of ^{131}I to Treat Cancer

Two methods of selecting treatment doses of ^{131}I are used. In one, the doses are chosen somewhat arbitrarily although the decisions are derived from experience with doses that appear to be effective and cause no serious side effects.[2,4,8] With this method, adult patients commonly receive about 150 mCi when the cancer resides in the neck and is not locally invasive, and doses of approximately 200 mCi for all other states in which the neoplasm is shown by scintigraphy to concentrate ^{131}I. The other method employs dosimetry of normal tissues and prescribes ^{131}I doses that, from experience, are just below the toxic range: when 120 mCi are retained in the body at 48 hours and less than 200 rads are imparted to the blood.[9,29] Doses up to 600 mCi have been prescribed by this method,[9] but, again, treatments are given only when the cancer is shown to sequester radioiodine.

Both methods evade the difficulty, often the impossibility, of applying dosimetry to thyroid cancer. Usually the carcinomas are too small or too ill-defined for estimates to be made of their volumes, and, without a tumor volume, no radiation dose can be calculated. Nevertheless, if dosimetry of the tumor becomes possible, it is incumbent upon the physician to calculate the radiation delivered since there is a threshold of tumor radiation below which ^{131}I is unlikely to be an effective therapy.[35]

We have usually employed the arbitrary doses of ^{131}I for patients in whom the tumor did not seem to pose any immediate risk to health or life. We applied dosimetry to patients in whom the cancer appeared to threaten health or life at the time or in the near future. Admittedly, it would be rational to adopt a plan based on the concept that if ^{131}I is worth giving as treatment, it should always be given by an optimal method, and that method now is the one based on dosimetry.

Summary

A physician who undertakes the treatment of well differentiated cancer must be well versed in the stages and the prognoses of patients with these neoplasms. Such knowledge will bring recognition that an oncologic team approach to these neoplasms is desirable. In addition, the potential and the hazards of ^{131}I treatments must be considered in any decision to use the radiopharmaceutical. With this background, only a patient whose thyroid cancer is likely to benefit from ^{131}I will be selected for scintigraphic study.

An optimal scintigraphic study demands proper preparation of the patient in terms of withdrawal from thyroid hormone and reduction of dietary iodine. Choosing the tracer dose of ^{131}I is somewhat arbitrary, but may be adapted to the urgency posed by the disease and probability of finding small cancer deposits. Scintigraphic images should be made 2 and possibly 3 days after the tracer dose to take advantage of the increasing tumor-to-background radioactivity. Care should be taken to survey the entire body of the patient and give additional attention to regions suspected of harboring metastases.

Whether and how to treat residual normal thyroid tissue remains controversial. There is no logic for treatment of small foci of normal tissue. Of the two methods of selecting ^{131}I doses to treat thyroid cancer, probably the use of dosimetry of normal tissue to calculate the highest tolerable dose is more logical (and more cumbersome) than the arbitrary choice of dose.

Discussion by Dr. Joel I. Hamburger

Dr. Sisson suggests that imaging at 5-year intervals may be appropriate in the follow-up of thyroid cancer patients after apparently successful ^{131}I therapy. I used to do that, but I never found anything, so I decided it wasn't worth putting the patients through the unpleasant experiences of thyroid withdrawal. Currently, I rely on examination, chest X ray, and a serum thyroglobulin assay. Is there any evidence that this is inadequate follow-up?

If low-iodine diet has the potential to increase retention of the ^{131}I, why not routinely force fluids and give diuretics beginning about 24 hours after the therapy?

Response by Dr. James C. Sisson

It may be that the thyroglobulin concentration will be reliable in predicting recurrences of well-differentiated thyroid cancers. However, the accuracy of this method over years is yet to be established. If levels of thyroglobulin reliably predict recurrent cancer, is there any need for a clinical examination or for a chest X ray?

Nevertheless, recurrences of thyroid cancer are seen.[4] New tumors appear in patients in whom the scintigraphic images over many years were negative. For example, one of my patients with papillary-follicular cancer, which was treated with thyroidectomy and ^{131}I in 1965 and was absent in multiple clinical examinations and images between 1966 and 1979, had a recurrence found clinically and on scintigraphic images in 1986. The tumor was then resistant to ^{131}I therapy. Whether annual thyroglobulin assays would have given earlier evidence of neoplastic resurgence cannot now be determined. However, the point is: these cancers do recur and we should adopt a reliable method for early detection of such recurrence.

Probably it was the use of diuretics rather than the low-iodine diet that increased the body retention of ^{131}I in the study of Maruca et al.[25] Urinary iodide excretion is related to the glomerular filtration rate. Unless diuretics have caused so much dehydration as to reduce glomerular filtration rate, a state that should be avoided, fluids will probably not alter the rate of excretion. As noted in my response to Dr. Wartofsky, the evidence supporting the efficacy of low-iodine diets is not very strong. Still, when prescribed for only a week, low-iodine diets will do no harm and may increase the uptake of ^{131}I by neoplastic tissue.

Discussion by Dr. John E. Freitas

I believe that the low-iodine diet increases the fractional uptake of ^{131}I by thyroid tissue with no significant increase in whole body radiation. When diuretics are added to a low-iodine diet, there is an increase in total body radiation but no significant increase in tumor radiation as compared to the low-iodine diet alone.[25,36]

Discussion by Dr. Leslie J. DeGroot

I take a different position in regard to ablation of residual thyroid tissue in patients without demonstrable metastases. Although I am up front about admitting that the treatment is of unknown efficacy, I think it has theoretical merit and no known risk. We tend to ablate postoperative residual thyroid tissue in most patients, except those who are young, perhaps under age 25, with single intrathyroidal lesions of under a centimeter. Certainly I would ablate patients who are older, past the age of 40, regardless of the size of the primary, who have nodes, who have larger tumors, or who have had a history of radiation exposure. Thus in our clinic most patients do receive postoperative radioactive iodide. The logic in ablation, of course, is that it may destroy foci of tumor in the thyroid if there is undetected multicentricity, which is common, or possibly elsewhere. On average it makes the TSH more elevated during subsequent periods of withdrawal of hormone. Ablation of the residue makes the thyroglobulin measurements, when the patients are off thyroid hormone, a test of unquestioned validity and clinical value. Residual functioning tissue in the neck, after one or two ablative doses, can be clearly identified as carcinoma.

The risk is that of approximately 10 or 12 rads whole body radiation, equivalent to background by age 30, and is, I believe, inconsequential. Whether ablation improves prognosis remains uncertain. We use a 30-mCi outpatient dose and five sixths of patients are ablated with one administration and the remainder with a second.

Response by Dr. James C. Sisson

I agree that ^{131}I treatment should be given to patients who have presumed or definite thyroid cancer in tissue that concentrates ^{131}I. I also agree that normal thyroid tissue should be ablated if the function of that tissue is sufficiently high to impair maximal secretion of endogenous TSH; in this state, any residual thyroid cancer cannot be detected or treated by ^{131}I in an optimal manner. Normal tissue is usually satisfactorily reduced for this purpose by a 30-mCi dose of ^{131}I.

However, I disagree that 30 mCi of ^{131}I is a reasonable dose to treat thyroid cancer. Our experience is that a number of well-differentiated cancers do not disappear even after a treatment with 200 mCi of ^{131}I. Some calculations also bolster my argument. In the presence of a normal thyroid gland, thyroid cancer will concentrate ^{131}I to a level that is $1/10$ or less that in the normal gland. This fraction was derived from data published on cancers prepared for and responding to ^{131}I treatment.[35] The neoplasm uptake is estimated to be the equivalent of 0.2%/g in contrast to normal tissue where uptake is about 1%/g. When the TSH concentration is not optimally elevated, the uptake in cancers is likely to be less than 0.2%/g, probably 0.1%/g or less. In this circumstance, treatment of the cancer with 30 mCi is the equivalent of 3 mCi (3,000 rads) or less for normal tissue. Such low doses usually fail to destroy the target; indeed more than 8,000 rads are necessary to destroy the majority of thyroid cancers.[35] If one wishes to treat thyroid cancer, the means ought to enable attainment of the end.

Whether all normal thyroid tissue should be eradicated is another controversy. One major goal of thyroglobulin concentrations will be the ability to detect change in the status of thyroid cancer. It has not been shown that an increment in the thyroglobulin from an undetectable level is more sensitive for this purpose than an increase from a measurable concentration. Therefore, the virtue of producing undetectable thyroglobulin values by eliminating normal thyroid cells with 30 mCi of ^{131}I has not been established.

When our surgeons leave thyroid tissue, the residual usually concentrates 2% or less of the tracer dose; this degree of function is unlikely to inhibit the secretion of TSH. If the surgeon indicates that he or she purposely preserved a small mount of thyroid tissue to protect the laryngeal nerve, and the pathologist does not find thyroid cancer (multifocal deposits or otherwise) near the region of preserved cells, I do not believe that ablation of such tissue serves a useful purpose.

Finally, I agree that ^{131}I treatments have been remarkably safe. Nevertheless, if we give small, or moderate, doses to many people, we increase the probability that some untoward event will occur. If radiation to the thymus had not been prescribed for many infants, we would today not know the hazards of such treatment. A risk, though low, should be worth the taking.

Follow-up Discussion by Dr. John E. Freitas

I, too would image a patient with regional or distant metastases even if the papillary primary was less than 1 cm. However, we do not usually image or treat irradiated patients with lesions <1 cm if no metastases are present.

We do not subscribe to the 30-mCi outpatient dose philosophy for residual thyroid bed uptake, but prefer to give 100 mCi to ablate these patients. Such ablation is essential if serum thyroglobulin plays an important role in your thyroid cancer follow-up program.

Discussion by Dr. Paul Lo Gerfo

Is there any evidence that "routine" postop scanning is of value to the patient?

Would you ablate a thyroid remnant (i.e., lobe) in someone with a microinvasive follicular lesion without evidence of mets from routine scanning?

Does scanning or treatment with ^{131}I have any role in a patient with Hurthle cell follicular cancer?

Response by Dr. James C. Sisson

Whether "routine" postop scanning is of value will depend upon the definition of "routine." For a patient with a "minimal" (less than 1 cm in diameter) papillary cancer found incidentally at the thyroidectomy, scintigraphy is not indicated. However, when the cancer is larger and especially when the risk of cancer to patient is appreciable, images made with ^{131}I are of value in both identifying the presence and the location of thyroid cancer. Occasionally, previously unsuspected distant metastases are uncovered by scintigraphy.[8,37] If one wishes to stage well-differentiated thyroid cancer, then scans will be an important component in the process.

If the thyroglobulin level were undetectable after thyroidectomy, then probably scans could be omitted. However, even the most skilled surgeon not infrequently leaves some normal thyroid tissue after a total thyroidectomy, and a measurable concentration of thyroglobulin will not distinguish between normal and neoplastic tissues.

We believe that normal tissue should be removed from patients who have cancers that pose defined risks. If a follicular cancer, even when small, exhibits vascular invasion, the risk to the patient is appreciable, particularly for distant metastases. In this circumstance, we would recommend removal of the remaining lobe to enable scintigraphy of the entire body. We prefer that the surgeon remove the lobe unless the risk of operation is too great. We have had too little experience in ablating whole lobes with ^{131}I to justify an opinion on the efficacy of this approach.

Most Hurthle cell carcinomas sequester little or no radioiodine.[38] However, within these neoplasms, there may reside follicular components that have an affinity for radioiodine. Since the treatment of residual Hurthle cell cancer has not been satisfactory, the small possibility that ^{131}I treatment would be effective may be worth exploring on images.

Discussion by Dr. Leonard Wartofsky

Do you believe that there is evidence of dose-related radioiodine damage to salivary glands? If so, shouldn't all patients undergoing periodic evaluations (including 2- to 10-mCi scans) and/or therapies be managed in a manner to minimize salivary irradiation?

How effective is a low-iodine diet ingested for only 1 week? Would a longer interval of reduced iodine intake coupled with iodine depletion (diuretics, salt) be of greater benefit (particularly in more aggressive tumors that do take up radioiodine)?

Please comment on the utility of thallium scanning for residual or recurrent thyroid cancer in your experience.

Do you obtain a scan image after therapeutic doses? If so, how do you follow up lesions seen on the posttreatment (large dose, 150 mCi) scan that were not visualized on the routine (2 to 10 mCi) neck and chest survey since one would not expect to see these lesions on a subsequent (2 to 10 mCi) survey either?

What is the evidence that low (30 mCi) doses of radioiodine will ablate normal tissue without damaging the ability of malignant tissue to take up radioiodine subsequently for therapy?

Response by Dr. James C. Sisson

Dr. Wartofsky has made a good point. We should always minimize the risks of what we do if it is reasonably possible. Thus, the radiation to salivary glands should be minimized even when tracer doses are given. The radiation from ^{131}I to the salivary glands can probably be reduced by increasing salivation via chewing gum or sucking on lemon drops. Radiation to the urinary bladder can be reduced by maintaining a state of hydration and frequent urination.

Iodine depletion can increase the uptake of ^{131}I by thyroid tissue in two ways. Barakat and Ingbar [23] showed that acute (within a day or two) iodine depletion caused a rise in the ^{131}I uptake by the normal thyroid gland under circumstances where TSH was unlikely to play a role; the mechanism appears to be one of autoregulation by thyroid cells. Acute iodine depletion is possible since there are only 90 μg in the human iodide pool while adults ingest more than 150 μg of iodide daily. The second way that uptake is enhanced is by reducing the substrate for hormone synthesis which, in turn, will bring increased stimulation of thyroid function by TSH; probably many days, and perhaps weeks, of iodine deficiency are required to evoke this mechanism. Thus, by prescribing a low-iodine diet for patients with thyroid cancer, we hope to bring out the first mechanism, the autoregulation of thyroid tissue, by diminishing levels of circulating iodide. Although it seems that well-differentiated thyroid cancers should retain this autoregulatory function, the evidence that a low-iodine diet in fact increases the uptake of ^{131}I in cancers is not substantiated. Because a low-iodine diet lacks both full nutrition and palatability, it is unlikely that patients will tolerate the diet for more than 5 to 7 days. Food containing less that 50 μg of iodine per day was associated with an increase in ^{131}I uptake by normal thyroid tissue but no change in the uptake by metastases.[24] When less than 25 μg of iodine per day were ingested for 4 days and combined with diuretics, there was an increase in tumor uptake of ^{131}I, but at the expense of prolonged retention of ^{131}I in the body.[25] In all studies, the number of patients evaluated have been few. Possibly the greatest advantage in adopting a low-iodine diet is to emphasize avoidance of high-iodine foods.

We have almost no experience in using ^{201}Tl to detect well-differentiated thyroid cancer. Hoefnagel et al[30] found ^{201}Tl images to be more sensitive but less specific than scintigraphy with ^{131}I for well-differentiated thyroid cancer. With the availability of reliable thyroglobulin assays, there will be little need for ^{201} Tl scans except to locate tumors that do not concentrate ^{131}I. For example, if a Hurthle cell cancer or medullary thyroid cancer is presumed to reside in the neck, the surgeon may be guided by evidence for functioning tissue observed in images made with ^{201}Tl.[39]

How does one measure the results of ^{131}I therapy when the only evidence of thyroid cancer is in the posttherapeutic images (i.e., to obtain the evidence a dose of 150 to 200 mCi of ^{131}I is required)? This question posed by Dr. Wartofsky is fundamental to treatment of thyroid cancer by ^{131}I. Asked, in another way, when should we prescribe treatment and when should we stop?

We have observed, in mock studies, that modern scintigraphic instruments cannot detect all treatable tumors (deposits that would receive 8,000 rads or more from 200 mCi of ^{131}I) even when the tracer dose is 30 mCi.[27] Based on this information an argument could be made for "prophylactic" treatment with ^{131}I. However, we believe that although some metastases may be undetectable by scintigraphy, these are usually associated with larger tumors that are visible on images. Therefore, the visible cancers serve as a guide to what is happening to the subliminal neoplasms.

Nevertheless, the question will be asked again when a patient has a rising concentration of thyroglobulin but exhibits no focus of radioactivity in scans made with 2, 10, or even 30 mCi of ^{131}I. Should such a patient receive therapeutic radioactive iodine? At this time, no one knows.

Thirty millicuries of radioactive iodine will often ablate normal tissue.[32,33] On the other hand, we do not have rigorous evidence that 30 mCi of ^{131}I will not impair the function of the coexisting malignant tissue. Probably in some cases, the uptake mechanisms of thyroid cancer will be affected by such doses. However, the 30-mCi doses should be prescribed only when: 1) the neoplastic tissue is not portrayed on image, and 2) the TSH concentration is normal or only modestly elevated. In this circumstance, the ability of thyroid cancer to sequester ^{131}I will usually be much below that of normal tissue, and the tumor radiation will then be similar to that after giving a 3-mCi tracer dose in a search for cancer in the absence of normal tissue.

An anecdotal case: 1 year after 200 mCi of ^{131}I were given to treat metastases of thyroid cancer, new functioning metastases were observed. In this patient, some neoplastic deposits, the function of which was initially undetected, eventually exhibited a substantial affinity for ^{131}I despite prior exposure to therapeutic radioiodine. The point of this case is: the function of the thyroid cancer was not unexpectedly sensitive to the radiation from ^{131}I.

There is another reason for using lower does of ^{131}I to ablate large quantities of normal thyroid gland. Functioning thyroid tissue will convert ^{131}I-iodide into ^{131}I-thyroxine, and the radioactive hormone circulates with a longer half-life than does radioiodide. When given to patients with substantial quantities of normal thyroid gland (of functioning neoplasm), larger doses of ^{131}I will result in appreciable radiation to blood and marrow; indeed, Leeper and Shimaoka warn that blood dosimetry is essential if one wishes to avoid bone marrow suppression in these patients.[9] Concern for the hazard of absorbed radiation is then another reason for limiting the dose of ^{131}I given for ablation.

Discussion by Dr. Ian D. Hay

On two major points, I am in entire agreement with Dr. Sisson. First, thyroid cancer should be treated by a "team of oncologic physicians," and second, scintigraphy should be restricted only to patients "whose thyroid cancer is likely to benefit from ^{131}I."Like Dr. Sisson, I consider that with our present knowledge of prognostic factors in papillary/follicular cancer, it is possible to classify patients, at the time of primary operation, into risk groups, which in the future may be helpful in deciding which patients should be more aggressively treated after primary surgery. We recently developed, from an extensive multivariate analysis, a prognostic scoring system for papillary cancer, based on *a*ge, *g*rade, *e*xtent, and *s*ize (AGES scheme). We believe that such a scoring system can define that minority of patients at highest risk of death from cancer [40] and could perhaps aid in deciding which patients might benefit from ^{131}I treatment.

Currently, we do not study or treat with ^{131}I those patients whose papillary tumors are <1-cm diameter, independent of nodal status at presentation. In a recent update of his U.S.A.F. cohort, Mazzaferri[41] continued to report that a reduction in tumor recurrence rates was not demonstrable in patients with primary tumors smaller than 1.5-cm diameter.

In relation to tumor surveillance, we have, like Dr. Sisson, had some experience with thallium-201 whole body scanning. However, to date, we have concerns regarding its specificity and sensitivity, and are not convinced that these images "will add perspective to the appraisal of cancer." Dr. Sisson's doubts regarding the reliability of serum Tg as an early warning device do not seem to be shared by Dr. Dunn in his scholarly analysis of the 1980–86 literature.

Response by Dr. James C. Sisson

Dr. Hay and his co-workers at Mayo Clinic have contributed greatly to our ability to assign relative risks to patients with well-differentiated thyroid cancer. Their scoring system using the independent factors of age, grade, extent, and size of papillary cancer in a patient will identify patients at high and low risk for death from this neoplasm.[40] With the exception of histologic grade, each factor can be readily applied to a patient; the method of histologic grading is not so easily adopted and may not meet with unanimous agreement among pathologists. Nevertheless, the practicing physician is searching for thresholds at which decisions are made for one type of treatment or another. Given the limited therapeutic options (extent of operation, large and small does of ^{131}I, and thyroid hormone), we should try to establish the threshold (the level on the risk scale) that will dictate the value of one or more treatments. To this end, Hay et al have demonstrated that conservative operations are as beneficial as total thyroidectomy even in patients in whom the risk of the cancer is high.[40]

When the primary tumor of papillary cancer is less than 1.5 cm, and especially when less than 1.0 cm in diameter, death and distant metastases are exceedingly rare (2,3). Such "minimal" cancers may or may not be associated with clinically evident regional node metastases. When the lymph node metastases are visible or palpable, the probability of clinical recurrence is 10-fold higher than when there are either no metastases or not clinically apparent metastases.[3] In fact, the risks of postoperative node involvement approach 1% per year over 5 to 10 years after thyroidectomy. Whereas these recurrences do not endanger life, they are frightening to patients. Most physicians would advise a reasonable prophylaxis against these new tumors.

Although Mazzaferri did not find that ^{131}I treatment reduced recurrence rates for papillary carcinomas of less than 1.5-cm diameter,[4] this is not the same as determining that ^{131}I therapy had no significant effect. As noted by Hay et al, type 2 statistical errors are difficult to exclude in analyses of modest numbers of patients.[40] More important, Mazzaferri did not examine separately those patients who presented initially with clinically apparent node metastases. This subgroup of patients who are at risk for recurrent cancer was diluted by patients in whom recurrence would not have been expected regardless of radioiodine treatment. Therefore, it is still possible that recurrences may be reduced by ^{131}I treatment to the patients who manifest small primary papillary carcinomas and obvious regional lymph nodes metastases.

Unfortunately, much of our treatment of thyroid cancer must be based on data that are incomplete. Thus, our reasoning will perforce be formulated from uncertain information. As does much of the practice of medicine, treatment of thyroid cancer rests on faith, hope, and charity (a charity that acknowledges that those who differ with us may hold some part of the truth).

Like all physicians, I hope that the thyroglobulin assay will be our early warning device for detecting recurrent and progressive thyroid cancer. As Dr. Dunn has carefully documented, thyroglobulin concentrations give promise of fulfilling our hopes. However, to date, studies have correlated the concentration of thyroglobulin with the prevalence of thyroid cancer. What we wish to know, and what will take time to learn, is how accurately and promptly changes in thyroglobulin levels predict recurrence or progression of disease.

Discussion by Dr. John T. Dunn

I am impressed with how unpleasant and disruptive it is for patients with postoperative thyroid cancer to become hypothyroid in order to be screened. In your patients at low risk for recurrence, is it really necessary, safe, or desirable to keep making them hypothyroid for ^{131}I scanning, a low-yield procedure for them? What is the yield on the every 5-year repeat?

Response by Dr. James C. Sisson

We do not know how many of our patients, at low to moderate risk for recurrent thyroid cancer, are found to have new tumors in periodic reassessments by scintigraphic images. The number is not large, but it probably is not insignificant. At the Mayo Clinic, the recurrence rate approached 1% per year for the first 10 years after thyroidectomy, and this rate was unrelated to size of primary and included new tumors in patients who had "minimal" carcinomas. The value of scintigraphies in discovering recurrent neoplasms is uncertain, but many physicians and patients would prefer to find them before they became clinically obvious. If thyroglobulin assays will serve as well, or better, to detect recurrent thyroid carcinomas in the face of fully suppressive doses of thyroxine, then images made with ^{131}I will be relegated to a secondary position in the periodic reevaluations of patients with thyroid cancer.

References

1. McConahey WM, Hay ID, Woolner LB, et al: Papillary thyroid cancer treated at the Mayo Clinic, 1946 through 1970: Initial manifestations, pathologic findings, therapy, and outcome. Mayo Clin Proc 1986;61:978.
2. Samaan NA, Maheshwari YK, Nader S, et al: Impact of therapy for differentiated carcinoma of the thyroid: An analysis of 706 cases. J Clin Endocrinol Metab 1983;56:1131.
3. Cady B, Cohn K, Rossi RL, et al: The effect of thyroid hormone administration upon survival in patients with differentiated thyroid carcinoma. Surgery 1983;94:978.
4. Mazzaferri EL, Young RL: Papillary thyroid carcinoma: A 10-year follow-up report of the impact of therapy in 576 patients. Am J Med 1981;70:511.
5. Young RL, Mazzaferri EL, Rahe AJ, et al: Pure follicular thyroid carcinoma: Impact of therapy in 214 patients. J Nucl Med 1980; 21:733.
6. Harness J, Thompson NW, McLeod MK et al: Follicular carcinoma of the thyroid gland: Trends and treatment. Surgery 1984;96:972.

7. Schlumberger M, Tubiana M, deVathaire F, et al: Long-term results of treatment of 283 patients with lung and bone metastases from differentiated thyroid carcinoma. J Clin Endocrinol Metab 1986;63:960.
8. Beierwaltes WH: Carcinoma of the thyroid—radionuclide diagnosis, therapy and follow-up. Clin Oncol 1986;5:23.
9. Leeper RD, Shimaoka K: Treatment of metastatic thyroid cancer. Clin Endocrinol Metab 1980;9:393.
10. Haynie TP, Beierwaltes WH: Hematologic changes observed following I-131 therapy for thyroid carcinoma. J Nucl Med 1963;4:85.
11. Sarkar SD, Beierwaltes WH, Gill SP: Subsequent fertility and birth histories of children and adolescents treated with ^{131}I for thyroid cancer. J Nucl Med 1976;17:460.
12. Benua RS, Cicale NR, Sonenberg M, et al: The relations of radioiodine dosimetry to results and complications in the treatment of metastatic thyroid cancer. Am J Roentgenol 1962;87:171.
13. Allweiss P, Braunstein GD, Katz A, et al: Sialadenitis following I-131 therapy for thyroid carcinoma: Concise communication. J Nucl Med 1984;25:755.
14. Brinckner H, Hansen HR, Anderson AP: Induction of leukaemia by ^{131}I treatment of thyroid carcinoma. Br J Cancer 1973;28:232.
15. Cronkite EP, Moloney W, Bond VP: Radiation leukemogenesis. Am J Med 1960;28:673.
16. Pochin EE: Long term hazards of radioiodine treatment of thyroid carcinoma. Hedinger CE, ed. Thyroid Cancer. New York: Springer-Verlag, 1969, pp 293–304.
17. Nel CJ, van Heerden JA, Charib H, et al: Anaplastic carcinoma of the thyroid: A clinico-pathologic study of 82 cases. Mayo Clin Proc 1985;60:51.
18. Goldman JM, Line BR, Aamodt RL, et al: Influence of triiodothyronine withdrawal time on ^{131}I uptake postthyroidectomy for thyroid cancer. J Clin Endocrinol Metab 1980;50:734.
19. Ramanna L, Waxman AD, Brachman MB, et al: Correlation of thyroglobulin measurements and radioiodine scans in the follow-up of patients with differentiated thyroid cancer. Cancer 1985;55:1525.
20. Girelli ME, Busnardo B, Amerio R: Critical evaluation of serum thyroglobulin (Tg) levels during thyroid hormone suppression therapy versus Tg levels after hormone withdrawal and total body scan: Results in 291 patients with thyroid cancer. Eur J Nucl Med 1986;11:333.
21. Hilts SV, Hellman J, Anderson J, et al: Serial TSH determination after T_3 withdrawal or thyroidectomy in the therapy of thyroid carcinoma. J Nucl Med 1979;20:928.
22. Melmed S, Harada A, Hershman JM, et al: Neutralizing antibodies to bovine thyrotropin in immunized patients with thyroid cancer. J Clin Endocrinol Metab 1980;51:358.
23. Barakat RM, Ingbar SH: The effect of acute iodine depletion on thyroid function in man. J Clin Invest 1965;44:1117.
24. Maxon HR, Thomas SR, Bochringer A, et al: Low iodine diet in I-131 ablation of thyroid remnants. Clin Nucl Med 1983;8:123.
25. Maruca J, Santner S, Miller K, et al: Prolonged iodine clearance with a depletion regimen for thyroid carcinoma: Concise communication. J Nucl Med 1984;25:1089.

26. Waxman A, Ramanna L, Chapman N, et al: The significance of I-131 scan dose in patients with thyroid cancer: Determination of ablation: Concise communication. J Nucl Med 1981;22:861.
27. Arnstein NB, Carey JE, Spaulding SA, et al: Determination of iodine-131 diagnostic dose for imaging metastatic thyroid cancer. J Nucl Med 1986;27:1764.
28. Henk JM, Hirtsman S, Gean CM: Whole body scanning and ^{131}I therapy in the management of thyroid carcinoma. Br J Radiol 1972;45:369.
29. Nostrand DV, Neutze J, Atkins F: Side effects of "rational dose" iodine-131 therapy for metastatic well-differentiated thyroid carcinoma. J Nucl Med 1986;27:1519.
30. Hoefnagel CA, Delprat CC, Marcuse HR, et al: Role of thallium-201 total-body scintigraphy in follow-up of thyroid carcinoma. J Nucl Med 1986;27:1854.
31. Beierwaltes WH, Rabbani R, Dmuchowski C, et al: An analysis of "ablation of thyroid remnants" with I-131 in 511 patients from 1947–1984: Experience at University of Michigan. J Nucl Med 1984;25:1287.
32. DeGroot L, Reilly M: Comparison of 30 and 50 mCi doses of iodine-131 for thyroid ablation. Ann Intern Med 1982;96:51.
33. Snyder J, Gorman C, Scanlon P: Thyroid remnant ablation: Questionable pursuit of an ill-defined goal. J Nucl Med 1983;24:659.
34. Sisson JC: Applying the radioactive eraser: I-131 to ablate normal thyroid tissue in patients from whom thyroid cancer has been resected. J Nucl Med 1983;24:743.
35. Maxon HR, Thomas SR, Hertzberg VS, et al: Relation between effective radiation dose and outcome of radioiodine therapy for thyroid cancer. N Engl J Med 1983;309:937.
36. Goslings BM: Effect of a low iodine diet on ^{131}I therapy in follicular thyroid carcinoma. J Endocrinol 1975;64:30.
37. Schlumberger M, de Vathaire F, Travagli JP, et al: Differentiated thyroid carcinoma in childhood: Long term follow-up of 72 patients. J Clin Endocrinol Metab 1987;65:1088.
38. Gundry SR, Burney RE, Thompson NW, et al: Total thyroidectomy for Hurthle cell neoplasm of the thyroid. Arch Surg 1983;118:529.
39. Arnstein NB, Juni JE, Sisson JC, et al: Recurrent medullary carcinoma of the thyroid demonstrated by thallium-201 scintigraphy. J Nucl Med 1986;27:1564.
40. Hay ID, Grant CS, Taylor WF, et al: Ipsilateral lobectomy versus bilateral lobar resection in papillary thyroid carcinoma: A retrospective analysis of surgical outcome using a novel prognostic scoring system. Surgery 1987;102:1088.
41. Mazzaferri EL: Papillary thyroid carcinoma: Factors influencing prognosis and current therapy. Semin Oncol 1987;14:315.

Section Summary: Noninvasive Evaluation of Structural Abnormalities of The Thyroid

JOEL I. HAMBURGER

Although the discussions on this subject were extended by some of the participants to include issues relative to the treatment of thyroid cancer, I shall confine my summary remarks to diagnostic methods, the focus of this volume.

What Is the Role of the Serum Thyroglobulin (Tg) Determination in the Follow-up of Thyroid Cancer Patients?

The test is most useful in patients who have had total ablation of the thyroid, surgical and with postoperative ^{131}I when necessary. Under those circumstances, Tg levels should be undetectable. Detectable values, and especially rising values, suggest the need for investigation for residual or recurrent thyroid cancer. For maximal benefit from Tg assays one must use a method sensitive to 1 ng/ml. Nichols Laboratory has such an assay commercially available. When Tg levels are less than 1 ng/ml, the chances of finding treatable thyroid cancer are too small to justify withdrawal from thyroid hormone and ^{131}I imaging in most cases. However, if there are overriding clinical considerations that increase the probabilities that there might be treatable cancer, one might still pursue ^{131}I imaging in spite of an undetectable Tg level. Also, it must be remembered that a positive Tg assay may reflect the presence of thyroid cancer that will not concentrate ^{131}I, and may thus be untreatable. Drs. Schlumberger and Tubiana have shown that imaging with 100 mCi of ^{131}I may demonstrate a low-level capacity to take up the tracer in these tumors. Under these circumstances the Tg assay would be a simpler way to monitor the course of the disease.

Is Thyroid Imaging Cost Effective in the Evaluation of Thyroid Nodule Patients?

The usefulness of thyroid imaging to differentiate thyroid nodules with increased function (virtually never malignant) from those with reduced function (malignant about 10% of the time) is well established. However, since less than 10% of nodules have increased function, and nodules with reduced function will be studied by needle biopsy to assess the risk of malignancy, some physicians have suggested that thyroid imaging is not cost effective for the diagnosis of thyroid nodules. One might better just perform needle biopsy as the initial procedure. Imaging can then be reserved for patients on whom needle biopsy diagnoses suggest follicular neoplasms that are not clearly malignant, or provide inadequate cellular material for reliable diagnoses. In sum this might constitute about one third of the patients. To save two thirds of the images would require biopsies on those 10% of nodules with increased function, for which biopsy would not have been needed had imaging been done. Since the cost of biopsy (including pathology charges) is about twice that of imaging where I practice, some of the savings would be illusory. I prefer to know what kind of nodule I am biopsying and I still use images to select those patients who do not need needle biopsy. Imaging is less threatening to most patients than needle biopsy, and this is another reason to restrict biopsies to those who really need them, even at the expense of extra imaging.

Is Thyroid Imaging Necessary Before Giving ^{131}I Therapy to Hyperthyroid Patients with Diffuse or Multinodular Goiters?

Freitas thinks imaging is useful to help differentiate between Graves' disease and toxic multinodular (autonomous) goiter. DeGroot does not think that information is important. I agree with DeGroot. It is true that larger doses of ^{131}I are desirable for hyperthyroid patients with multimodular goiters, but this is true whether the patient has Graves' disease with incidental nodules, or the autonomous tissue of toxic multinodular goiter. Palpation along with ^{131}I uptake determination is adequate to make a judgment on dosimetry. Nevertheless, I do perform a thyroid image before ^{131}I therapy to provide an objective documentation of the structural status of the thyroid gland before I treat it. These images have been useful to me on numerous occasions when patients subsequently claimed that they had not needed the treatment, or did not benefit from it.

Which Thyroid Cancer Patients Should Be Studied Postoperatively by Imaging for Possible ^{131}I Ablative Therapy?

One should study only those patients for whom one anticipates a possible benefit from ^{131}I therapy. To some physicians this means all patients with differentiated thyroid cancer. DeGroot provides the following rationale for total ablation of thyroid tissue in most thyroid cancer patients (using first surgery and then ^{131}I). This treatment may destroy tumor foci that might exist in the residual tissue but were not appreciated at the operation, or even tumor elsewhere. It assures that TSH levels will be optimally elevated on subsequent studies after withdrawal of thyroid hormone. Ablation of all thyroid tissue enhances the usefulness of the Tg assay. After full ablation of normal thyroid tissue by surgery and one or two doses of ^{131}I, the subsequent detection of tissue in the neck that concentrates ^{131}I is unequivocal evidence of carcinoma. Finally, ^{131}I therapy carries no appreciable risk.

These are persuasive arguments, and I tend to advise ^{131}I therapy rather liberally as well. However, I do not think that low-risk patients have enough likelihood of benefit from ^{131}I therapy to justify the unpleasantness of hypothyroidism and the expense of imaging procedures. Having said this, there is only the need to define "low risk." DeGroot spares patients younger than 25 with solitary intrathyroidal lesions smaller than 1 cm from ^{131}I therapy. Because I, like Freitas, am impressed with the reports of a higher prevalence of pulmonary metastases in young people with thyroid cancer, and because I agree with Freitas that there is a better chance of eliminating lung metastases when they are detected by imaging before they reach the size at which they are obvious on chest roentgenogram, I would advise imaging on young patients, but would not treat a small remnant within the thyroid bed. For patients 25 to 40 years old with solitary lesions 1.5 cm or smaller, and no evidence of metastatic disease, I would not advise imaging because it is too unlikely that ^{131}I therapy will be beneficial. I would make exceptions, however, for any who have an unexpected elevation in Tg concentration.

Should Patients Be Placed on Low-Iodine Diets in Preparation for Postoperative Imaging and Possible ^{131}I Therapy?

Discussion of the use of low-iodine diets in the preparation of thyroid cancer patients for postoperative imaging and ^{131}I therapy was rather subdued. Sisson and Freitas advised them, but Sisson's endorsement was less than enthusiastic, saying that the "greatest advantage in adopting a low-iodine diet is to emphasize avoidance of high-iodine foods." As the one who first introduced the concept of stable iodine depletion to augment ^{131}I uptake by inoperable thyroid cancer,[1,2] perhaps I have a proprietary interest in this subject. There is ample evidence in the literature that low-iodine diets do increase uptake of ^{131}I by thyroid cancer. It is also widely appreciated that the best opportunity for effective ablation of inoperative thyroid cancer with ^{131}I therapy is the first time the treatment is given, because incomplete ablation after the initial ^{131}I dose will be associated with a poorer uptake of ^{131}I subsequently. Hence maximizing uptake of ^{131}I in the residual cancer prior to initiating ^{131}I therapy would seem to be such an obvious thing to do it is hard to understand why everyone does not use a low-iodine diet.

Should Thyroid Cancer Patients Be Imaged 7 to 14 Days After ^{131}I Therapy?

There is agreement by those who have tried it that this technique will reveal foci of ^{131}I uptake that are not seen after conventional tracer doses of ^{131}I. Wartofsky has tried it, but does not know how the information contributes to the patient's treatment. DeGroot tried it, found the information confusing and not useful, and no longer advises the study. Sisson suggests that the detection of new foci of ^{131}I uptake on images performed after a therapeutic dose of ^{131}I may have value from the prognostic standpoint, and also to indicate the need for closer follow-up, especially if there is uptake in the spine or lungs. Dunn seems to indicate that he does not advise it, and offers the opinions of Harry Maxon and David Becker in support of his attitude that tumor foci only visible after ± 150-mCi "tracer" doses would not be treatable with ^{131}I. I have never done these studies, following the dictum that ignorance is bliss in this situation. In 28 years of dealing with more than 1,000 patients with differentiated thyroid cancer, I have detected no adverse consequences from failing to obtain images after ^{131}I therapy. The evidence of potential benefit offered by others is anecdotal at best. Although I can see some possible merit in the investigation of this technique, it should be considered investigational until hard data are available to show that it is justifiable.

How Often Should Thyroid Cancer Patients Be Withdrawn from Thyroid Hormone and Subjected to Imaging in the Search for Persistent or Recurrent Thyroid Cancer After Ablation, Surgical and When Necessary with ^{131}I?

Sisson advocates imaging at 5-year intervals, and describes one case of a recurrence detected 20 years after apparently successful ablation. Schlumberger and Tubiana image annually for 2 to 3 years. If those studies show that tracer uptake has been ablated, further follow-up is by clinical evaluation, with further imaging confined to those with elevated Tg levels. The other panelists were notably silent on this subject. Hence I feel free to present my own attitude on the follow-up of thyroid cancer patients. Following ^{131}I ablative therapy, I routinely image 1 year later. If the residual tissue for which ^{131}I had been given is no longer visible, and there are no new foci of ^{131}I uptake, I resume thyroxine in a dosage adequate to suppress TSH and reevaluate annually. My reevaluation includes physical exam, thyroid function tests to assure that TSH is well suppressed, a serum Tg level, a serum calcium, and a chest X ray.

In recent years since I have employed a low-iodine diet prior to ^{131}I therapy, I have had only one patient whom I can recall who failed to have ablation of all residual tissue. That patient had only a tiny residual focus of uptake remaining within the thyroid bed with a ^{131}I uptake of less than 1%, and I advised observation, rather than further ^{131}I therapy. I use only a 500-μCi tracer dose, and this may be part of the reason I almost never see persistent foci of ^{131}I uptake. However, it has long been by belief that treatable tumors will be seen with even a 500 μCi tracer dose of ^{131}I. I have seen the reports indicating the need to use 1 mCi-, 10 mCi- and even 30-mCi tracer doses, but I have not found them convincing for the simple reason that hundreds of patients who had differentiated thyroid cancer return to my office annually for reevaluation and almost never do I find evidence of recurrent disease in those for whom ablation has been achieved (ablation being defined as no abnormal uptake of ^{131}I after a 500-μCi tracer dose). Most of the recurrences I have seen have been in the neck, and were easily excised. I have had so few deaths from differentiated thyroid cancer that I cannot see any justification for more intensive follow-up. Those who have died have nearly always had untreatable local or metastatic disease when I first saw them.

My approach to the follow-up of thyroid cancer patients is influenced heavily by the population of thyroid cancer patients I serve. These are predominantly patients with smaller tumors detected by primary care physicians on routine exams. These tumors are representative of the unselected disease as it will present in most medical facilities. In university centers where the highly selected most aggressive cancers are more frequently seen, the more aggressive treatment and follow-up practiced by Sisson and DeGroot may well be appropriate.

References

1. Hamburger JI, Desai T: Mannitol augmentation of ^{131}I uptake in the treatment of thyroid carcinoma. Metabolism 1966;15:1055–1058.
2. Hamburger JI: Diuretic augmentation of ^{131}I uptake in inoperable thyroid cancer. N Engl J Med 1969;280:1091.

Section III
Needle Biopsy Diagnosis of Thyroid Nodules

9

Coarse-Needle Biopsy of the Thyroid

PAUL LO GERFO

Needle biopsy of the thyroid gland has almost become routine in the management of thyroid nodules. Fine-needle biopsy (FNB) has become widely accepted by surgeons and internists as a relatively low-risk procedure that is extremely useful.[1]; Coarse-needle biopsy (CNB) (Tru Cut, Vim Silverman) has not gained wide acceptance, but appreciation of the value of this biopsy technique appears to be increasing in the surgical community.[1-3] Despite the extensive use of needle biopsy, there is little evidence to show that it has improved our ability to select patients who have a higher incidence of cancer at the time of thyroidectomy.[4] In fact, there is some evidence to suggest that the FNB has resulted in an incidence of cancer in thyroidectomy specimens that is lower than that before its extensive use. For example, Ashcraft and Van Herle reported a prevalence of cancer of 20% in thyroid nodule patients operated upon on the basis of FNB findings.[1] However, without FNB study, Hill reported a 36% prevalence of cancer in patients operated upon after a trial of thyroid hormone therapy failed to produce nodular regression.[4] Regardless of techniques used, probably the most important factor in selecting patients is the physician's experience and knowledge about thyroid disease, including pathology. It would be difficult for most people to gain the experience most of us have. For example, it is ridiculous to expect anyone to become proficient at coarse-needle biopsy if he or she is only seeing 20 to 30 new nodules a year.

For the clinician, most of the problem in understanding the controversy surrounding thyroid biopsy techniques relates to incomplete understanding of thyroid pathology. It is surprising how little most clinicians know about pathology and how little most pathologists know about clinical thyroidology. In any event, for those interested in thyroid disease I believe a solid base in all areas of thyroidology are essential. The reason the cancer incidence at surgery cannot increase much beyond 50% is that some thyroid lesions, termed microfollicular adenomas, are difficult to distinguish from follicular cancers without examining the entire lesion for venous or capsular invasion. The important word here is microfol-

TABLE 9.1. Incidence of thyroid cancers and microfollicular adenomas in 748 patients.

Cancer	%
Papillary (mixed)	42.8
Follicular (including microfollicular)	9.7
Medullary	2.8
Anaplastic	1.7
Microfollicular adenomas (benign)	42.8
Overall cancer incidence	57.2

licular, not adenoma. The latter is a very loose term that is applied to a wide variety of lesions. In our experience only about 19% of these microfollicular lesions are cancer, but unfortunately this distinction can only be made after removal. The practical significance of this is that the incidence of cancer in any series is going to be diluted by benign microfollicular lesions. Table 9.1 illustrates our experience with this problem.

These data, representing 10 years of experience with thyroid cancers and microfollicular adenomas at Columbia Presbyterian Hospital, have established the ratio of microfollicular adenomas to follicular cancer. Assuming all cancers are removed and all microfollicular lesions are removed (9.7% are included with follicular cancers), then the maximum incidence of cancer in surgical specimens by the physician who removes all lesions at risk would be approximately 57%. The problem is further complicated by the fact that adenomatous hyperplasia (macrofollicular lesions) is an extremely common condition that can be confused with microfollicular adenomas (Table 9.2). Distinguishing these benign lesions from potentially malignant microfollicular lesions is easier with more tissue and can more readily be done with larger aspiration needles or cutting needles. It cannot be done with fine-needle aspiration biopsy.

The technique of FNB is well described by Dr. Hamburger and there is little need for me to elaborate on this. I would, however, point out that I prefer to use a larger needle for FNB (i.e., 18 gauge). These larger needles give the better structural detail that is difficult or impossible to get from 25 gauge or even 22 gauge needles. These larger needles will easily allow the pathologist to identify the microfollicular lesion since clumps of microfollicules will come through the needle. Also in some papillary lesions it is easier to distinguish the benign hyperplasias from papillary carci-

TABLE 9.2. Incidence of follicular cancer in macro versus microfollicular lesions.

	Number	Incidence of cancer
Microfollicular	80	21%
Macrofollicular	97	0%

nomas if nuclear cleaving is absent. If there is an increased risk from these larger aspiration needles, it is minimal. Certainly carotid and jugular punctures are commonly done for vascular access with little or no morbidity, and it is rare that the thyroid is more vascular than those structures. I have found that these larger needles are almost as good as coarse-needle biopsies and I use them regularly in patients with nodules too small for CNB.

The technique of coarse-needle biopsy is definitely more difficult to learn and requires much more experience to become truly proficient. I have often spoken with physicians who told me that CNB is too risky and that one cannot get reliable specimens. It is rare that any of those physicians have tried the procedure more than 10 times. Any physician who is not doing more than 50 of these procedures per year is unlikely to ever get very good at doing them. At a recent meeting two fellow surgeons got up to speak about the dangers of CNB. Together they had attempted CNB three times.

The technique of CNB is similar to FNB. The patient is placed in a supine position with head extended. Occasionally, I place a pillow under the shoulders. The tissues in the midline of the neck above the lesion are injected with xylocaine (I prefer epinephrine) and the injection is carried down to the lesion. I generally aspirate the lesion at this point to see if it is cystic or solid. If it is a cyst, I do not withdraw fluid because it is easier to biopsy the wall of a cyst if it is filled with fluid. If it is solid but very vascular by aspiration, it is almost certainly a follicular carcinoma or a vascular adenoma. For about 3% of the extremely vascular lesions I do not do CNB, but prefer to do an 18-gauge FNB. Occasionally, a calcified or gritty lesion is encountered. The gritty lesions are almost always papillary carcinomas. Calcified lesions, which are rare, may be too hard to penetrate with a Tru Cut needle and are generally benign. When doing the biopsy, it is important to insert the needle near the midline rather than through the sternocleidomastoid muscle, because biopsies done through the sternocleidomastoid muscle are associated with much more pain. A small incision with a number 11 blade is made through the skin and platysma. I prefer Tru Cut needles (3-inch) for the actual biopsy. These are extremely sharp and make the biopsy much easier than other needles I have tried. The capsule of the lesion is penetrated and the inner shaft of the needle is advanced and then withdrawn slightly to catch the posterior edge of the slot on the capsule of the lesion. It is easy to feel this if upward pressure is maintained on the tip of the needle. The hand controlling the central portion of the needle is rested on the chin or side of the face and the cutting portion is pushed in and the needle removed in closed position. Upward pressure is maintained on the needle to ensure that tissue remains in the specimen slot during this maneuver. The most common error is not fixing the hand holding the central portion of the needle when taking the biopsy. If this is not done, there is a strong tendency to move both hands and this simply does not yield good specimens. Upward pressure is also extremely important on the softer colloid lesions. The wall of a cyst is biopsied in exactly the same manner. I generally obtain three specimens. The entire procedure from lying the patient down to finish is approximately 5 minutes. If there is any bleeding, pressure is applied for a brief period.

I generally inspect the specimen with a hand lens (2×) to ensure that a capsule is present and to determine if the lesion is benign. Sixty percent of the time a definitive benign diagnosis can be made with just this inspection. The rest of the time it is necessary to await microscopic examination. Students who have spent time with me as preceptees have often initially doubted the reliability of these diagnoses. After reviewing the pathology findings and correlating them with the gross appearances, most of them are capable of predicting final diagnoses in a significant number of cases (e.g., fibrous, colloid, hyperplastic areas, Hurthle change, and degeneration all can be grossly recognized). It is important to correlate gross appearance with the final pathology diagnosis to be sure the lesion was biopsied. Capsules, when present, provide documentation that the lesion was biopsied. This is particularly important in the smaller, less accessible lesions.

It is difficult to biopsy lesions smaller than 1.5 cm, and for lesions of this size I use aspiration biopsy with 18-gauge or 20-gauge needles. In early studies approximately 25% of nodules seen were too small to biopsy. Current figures are slightly lower (20%) but this obviously remains a problem that can be satisfactorily handled with aspiration biopsies. Smaller nodules, particularly in the isthmus, I generally excise in the office under local if there is some doubt following aspiration biopsy.

Cysts in the thyroid are rarely malignant but there are reports indicating that the incidence may be as high as 7%. Certainly in my own experience the incidence of malignant cysts is less than 0.5%, but I generally biopsy the wall of these lesions to be sure. The ability to do this requires some practice, but good specimens are not difficult to obtain after going through a learning curve. Aspiration cytologic on some of these degenerated hyperplastic lesions can be extremely bothersome and a CNB in this situation will often avoid surgical intervention. The main risk with this procedure is bleeding and I have had two patients (of more that 3,000) on whom I operated within 24 hours of their initial biopsy. One of these had follicular carcinoma and one had nontoxic nodular goiter. This problem, although important, must be viewed in the context of the many unnecessary operations avoided.

We have carried out several studies comparing the incidence of cancer in surgically removed specimens based on FNB and CNB results in the same patients (Table 9.3). These data show that CNB yields better results than FNB. These findings are supported by others.[3,5–7] Results of a similar study comparing CNB and FNB to clinical methods prior to their intro-

TABLE 9.3. Results of biopsy.

Results	FNB	CNB
Cancer or suspicious	28	23
Adenoma	53	34
Thyroiditis	11	15
Benign	208	228
Number recommended for surgery	81	58
Incidence of cancer at surgery	33%	45%

duction at Columbia are seen in Table 9.4. As with previous studies, only CNB yielded better results than clinical methods. The problem, of course, is not whether one procedure is better than another, but understanding the limitations of each technique and knowing which one to use. I see many patients who would do just as well with only a FNB. CNB could easily be reserved for patients with borderline results. Our current practice of using 18-gauge needles for aspiration biopsy may in fact be almost as good as CNB, and we are now evaluating this. The controversy between FNB and CNB therefore is not which yields better results but whether the pain and risk of CNB is worth the extra benefit. Patients who want to avoid surgery would probably say yes. Surgery, of course, has become simpler for the patient and it is unusual that hospitalization is over 24 hours. In my current practice many patients are selecting outpatient surgery and some of these are being done under local anesthesia. Nevertheless, the risk of surgery is greater than CNB.

It has been said that microfollicular lesions of the Hurthle cell type have a greater risk of vascular or capsular invasion (i.e., cancer) than the non-Hurthle cell type. Table 9.5 gives our experience, showing that the incidence of invasion is the same whether the lesion is a Hurthle cell variant or not. The problem confronting the cytologist is that Hurthle cells can look very malignant at the cellular level, but structural relations are more important in defining the probability of malignancy for these lesions. CNB is definitely better in providing this information.

Dr. Wang (Massachusetts General Hospital), who has much experience with CNB, has had results similar to my own.[7] CNB will never become

TABLE 9.4. Incidence of cancer in surgical specimens before and after introduction of needle biopsy.

	No. of pts.	% Malignancy
Before introduction of needle biopsy at CPMC*	100	28
FNB at CPMC*	400	27
CNB at CPMC*	122	46
Needle biopsy dependent on nodule size at CPMC*	500	41

* CPMC = Columbia Presbyterian Medical Center.

TABLE 9.5. Incidence of carcinoma (CA) in follicular neoplasms.

Total number of follicular neoplasms	390
Microfollicular adenoma	252
Hurthle cell adenoma (pure)	71
Follicular CA (microfollicular)	53
Hurthle cell CA	14
Follicular CA 17% of all follicular lesions	
Hurthle cell CA 19% of all Hurthle cell lesions	

extensively used because of the risk involved in the hands of the uninitiated physician. Other theoretical risks do exist but have not been reported by experienced clinicians. I believe that one of the reasons CNB is not more widely used is that nonsurgeons are afraid to do the procedure because they cannot handle the complications. Most surgeons gain little experience because they are not involved in selecting thyroid nodule patients for surgery, and once a case is referred they have scant motivation to do a CNB to save the patient a surgical procedure. I recently reviewed my experience on patients who came to me after they were told they required surgery based on a FNB. After reviewing the cytology in these patients I recommended CNB for 212 of them and only 97 underwent surgery based on their CNB results. Admittedly this is a very selected group, but it does point out that FNB as it is being done by many endocrinologists is not as selective as CNB.

Most of my patients are referred to me after having undergone radioactive scans, ultrasounds, CAT scans, and other studies. I really do not see a need for most of these, and the cost in terms of personal anguish, lost work, and expense is extremely high. Most patients do not need any workup other than physical exams and biopsies of their nodules. Thyroid scans should only be done if a hot nodule is suspected (i.e., suppressed contralateral lobe or occasionally clinical symptoms). Ultrasounds add little other than to document what is found clinically. Most of these tests are treating the physician rather than the patient but most physicians continue to get them routinely. I do get thyroid function tests to document the baseline status of the patient, since many are eventually given thyroid hormone. From a practical standpoint all we need do is to make sure they are euthyroid and rule out cancer. Needle biopsy and thyroid function tests will do this for most patients.

Needle biopsy can be more accurate than frozen-section diagnosis particularly when dealing with papillary cancer. Even FNB is extremely accurate in some situations (papillary) and frozen section is not necessary. This will sometimes save time at surgery and allow one to explain the surgical plan to the patient preoperatively. I still believe that intraoperative judgments are extremely important. For instance, if the contralateral lobe is atrophic and small without nodules, multicentric disease or intrathyroid lymphatic spread is unlikely. The anatomy of the parathyroid gland and the appearance of the removed lobe (looking for multicentricity) are extremely important. Evidence of nodal metastases and the age of the patient affect intraoperative decisions.

In every single series of papers on FNB the percentage of people with thyroid nodules who are sent for surgery is higher than the percentage sent to surgery who have undergone CNB. This includes data from Hamburger who sends approximately 25% of his patients for surgery (most series are 20%, Table 9.1). In people who have undergone CNB the percentage is 16% or less. Our number, based on 3,000 patients, is 15%. It is also apparent from the literature that at a time when all thyroid nodules were excised the incidence of cancer was only 3% to 4%.[4] It should be apparent to the most naive reader that if we are dealing with patient populations with the same prevalence of cancer, the larger the proportion of patients having surgery, the lower the frequency of cancer will be at surgery. Stated in another way, the goal is to decrease the number of patients sent to surgery by better selection. This requires that a smaller percentage of patients should be sent to surgery. A larger proportion of patients studied by FNB have surgery than those studied by CNB. This has been attributed to differences in nodule size and that is why we did comparative series on the same patients. Therefore, until people advocating FNB can approach the low incidence of patients sent to surgery that CNB produces, it is difficult for me to embrace FNB except when necessary.

In summary, the use of CNB will result in lower percentages of patients being sent to surgery and a higher incidence of cancer at surgery. Both of these factors must happen together. In any series in which the number of patients referred to surgery exceeds 16%, the value of the procedure is not as good as CNB, assuming a constant false negative rate. A continuing problem is separating microfollicular lesions from hyperplastic ones, and separating both of these from cancers.

Discussion by Dr. Joel I. Hamburger

Dr. Lo Gerfo says there is little evidence that the use of FNB has led to a higher prevalence of cancer in surgical specimens. He supports that contention with the 1981 collected data of Ashcraft and Van Herle. But those data are not really representative of what FNB can do. In our hands FNB quickly led to a 50% reduction in the proportion of nodules for which surgery was advised, and a doubling of the prevalence of cancer in surgical specimens. Between 1985 and 1987, only 158 (18%) of the 888 nodules we studied by FNB were excised.[8] The prevalence of malignancy in surgical specimens was 44%, 56% if we exclude the 33 patients who had surgery in spite of FNB diagnoses of benign disease. (Curiously this 56% figure is very close to the 57% in Dr. Lo Gerfo's Table 9.1). Of course these FNB results were achieved only after a basic training period of nearly 10 years with more than 5,000 patients. Nevertheless, we believe that others can reach this level of FNB performance more quickly if they make use of published didactic and illustrative material that emphasizes avoidance of pitfalls.

I second Dr. Lo Gerfo's advise that it is ridiculous to expect proficiency in CNB or FNB if only 20 to 30 new nodule patients are seen annually. Indeed much of the misleading poor results with needle biopsy comes from centers reporting on only a few hundred cases or less.

I have used 18- and 16-gauge needles only to aspirate tissue fragments for histologic evaluation. I use the 25-gauge needle for FNB, and with an in-and-out movement seem to produce enough tissue disruption to provide small tissue fragments that our cytopathologists seem to find adequate for diagnoses. However, follicular adenomas are often not purely micro- or macrofollicular, but rather composed of a mixture of those patterns, especially the larger follicular lesions. It is because follicular neoplasms, especially the larger tumors for which CNB might be suitable, are so often pleomorphic that I favor FNB. With FNB I can sample multiple sites within the tumor more simply than with CNB. A representative case had CNB diagnoses of benign nodular goiter, follicular adenoma, and follicular carcinoma made on three separate occasions 1 to 2 years apart. The FNB was consistently indicative of a suspicious follicular neoplasm. The surgical specimen was a highly pleomorphic, follicular carcinoma with part of the tumor showing histologic patterns consistent with all three CNB specimens.

Having said this, I would add that discussions I have had with many endocrinologists indicate undue timidity with FNB. There is great reluctance to take more than two or three FNB samples. I consider this inadequate sampling, and compared to that use of FNB, as Dr. Lo Gerfo said, CNB is bound to be more reliable.

I agree with Dr. Lo Gerfo that Hurthle cell tumors seem to be malignant about as often as microfollicular (or mixed micro-macrofollicular) tumors. However, I wonder if he agrees that the Hurthle cell cancer tends to be more aggressive than the follicular carcinoma? Also, the metastases of Hurthle cell cancer do not concentrate ^{131}I, in contrast to follicular carcinomas. Hence, early surgery for Hurthle cell tumors is important because there is no effective fall-back therapy.

I think that Dr. Lo Gerfo has achieved an admirable level of success and accuracy with CNB and the safety and simplicity in his hands are equally commendable. I also agree with him that those who cannot approximate his good CNB results with FNB should seek further evaluation before committing the patient to surgery. However, it is possible to achieve comparable results with FNB, and the simplicity and safety of FNB, even in the hands of those with limited experience, are important advantages. Hence, I think that if effort is to be expended, it should be toward improving FNB results, rather than learning CNB (unless the volume of work justifies learning both). My own experience has led me to use CNB less and less often, so that now less than 5% of my patients have it. Nevertheless, it is hard to be sure what the final word will be on this subject. The big question is how reliably can needle biopsy (FNB or CNB) exclude cancer in tumors with any sizeable microfollicular component? In our zeal to avoid unnecessary surgery, how great is the danger of failing to identify low-grade follicular carcinomas, and what risks does that present to our patients? Good FNB technique leads to a reduction in the proportion of our nodule patients having surgery from 50% (before needle biopsy) to about 15% and an increase in prevalence of cancer in surgical specimens from 25% to more than 50%. Is this not good enough, or should we try to do still better by applying CNB to the follicular neoplasm whenever possible? Since suspicious follicular neoplasm (including Hurthle cell tumors) constituted only about 6% of the cases in my most recent series, it is easy to see why I don't think that using CNB can have much of an impact on the overall results. Dr. Lo Gerfo's experience is different in this regard, probably because he sees more patients for reevaluation after those with the easy FNB diagnoses have been treated. Thus his approach is no doubt correct for his patient population. What is needed for others will depend on the mix of patients they see.

Gharib et al[9] reported that 20% of thyroid nodules treated with a placebo shrank more than 50% in 6 months. Do you find these data believable? In my practice 8% of untreated nodules exhibited such a reduction, compared to 29% of thyroxine-treated nodules.

Response by Dr. Paul Lo Gerfo

Dr. Hamburger has published a series that shows that 25% of his patients are going to surgery. All other FNB groups are sending only 20% of their patients to surgery. In the CNB groups the number sent to surgery is 15%. I cannot reconcile the fact that Dr. Hamburger is sending more patients to surgery and is still achieving a higher cancer incidence at surgery. These results are contradictory (i.e., if a greater percent of patients are sent to surgery then the incidence of cancer at the time of surgery must be less). The only way to explain these findings is to assume all the rest of us are following a lot of undiagnosed cancer, or that our populations are entirely different than his. In addition it should be pointed out that all other series (including ones in which all nodules were excised) report a cancer incidence in their patient populations of 3% to 6%. Dr. Hamburger reports an incidence of more than 10%. These numbers are so different from the rest of the world it is difficult for me to understand. It also should be pointed out that we have done comparison studies and would suggest Dr. Hamburger do the same. He might even do better with CNB than we do.

The more tissue the better the diagnosis—I believe we all agree on this. Twenty-two-gauge needles give less than 18 gauge and less than CNB.

The ratio of microfollicular lesions to follicular cancer is constant. This ratio is 5:1. If a series shows 20% follicular cancer (among cancer types), then five times as many microfollicular lesions must be removed to find these cancers. All series report the incidence of follicular cancer as 12% to 22% of all cancers, therefore I do not know where this figure comes from. I would also like to point out that I like operating on patients and in addition I get paid for it. FNBs are also easier for me to do. My use of CNB therefore and my interest in CNB is strictly pragmatic. It gives the best results. I would like to point out that every single investigator who has taken the time to do comparison studies between FNB and CNB has reached the same conclusion as I have. I would encourage all of you do to this, including Dr. Hamburger.

I am currently reviewing some data but my impression is that if the patient has a hyperplastic lesion or one without much scarring or degeneration, the lesion has a good chance of decreasing in size. These lesions are determined by CNB. In general these tend to be in younger patients.

Discussion by Dr. Joel I. Hamburger

The 25% surgical referral for thyroid nodule patients is ancient history. Currently we send only 15% of our patients to surgery, essentially the same as Dr. Lo Gerfo. I cannot account fully for the higher cancer prevalence in our nodule patients. It was 7.9% in the 1985–87 series. Part of the difference may be that the number relates only to solid, cold nodules, not all nodules. Also we commonly do FNB on nodules so small that others have refused to try biopsy. For example, of the 59 papillary carcinomas we studied by FNB, 16 were only 1 cm in diameter, 5 were 3/4 cm, and 1 was 1/2 cm. If the nodules had not had biopsy evidence for cancer, most would have been observed because of their small size. Surely they were too small for CNB.

Larger tissue samples of CNB may not permit better diagnoses in large follicular tumors where pleomorphism is so common. Sampling of multiple sites by 8 to 10 FNB aspirates may actually provide more reliable diagnoses.

Dr. Lo Gerfo is correct that most series of thyroid cancer show a prevalence of follicular cancer of 12% to 22%. Our prevalence of 6% is substantially less. I believe the 6% figure for follicular carcinoma and 84% for papillary carcinoma more closely reflect the actual prevalence of the two diseases. We include the follicular variant of papillary carcinoma in the papillary group, even though the tumors may be predominantly follicular. Also, as already noted, we are diagnosing many small papillary carcinomas that others may be observing.

Finally, we have done comparison studies of CNB and FNB early in our experience. It is because CNB offered no advantage to FNB for nearly all of our patients that we have gradually phased out CNB. It may not be appropriate to compare FNB with CNB until one has reached the highest level of accuracy with FNB. Once this has been achieved, there is little incentive to use CNB. Nevertheless, this is a decision that each physician must make, depending upon the mix of nodules he or she encounters and the expertise of his or her cytopathologist.

Discussion by Dr. John E. Freitas

Experience in our institution and the work of Kini, Hamburger, Miller, and others demonstrates conclusively that fine-needle aspiration decreases the number of thyroidectomies performed for nodular disease, yet increases the yield of thyroid cancers.[10,11]

Response by Dr. Paul Lo Gerfo

This of course is true if the cancer incidence was low before FNB was introduced, and this was the case at most institutions. In addition, doing FNBs by a select group like Hamburger will lead to better patient selection no matter what methods are used.

Discussion by Dr. Ian D. Hay

Dr. Lo Gerfo states that FNB is not as selective as CNB, and cites data to show that after FNB 20% to 25% are sent for surgery, in contrast to a figure of 16% or less for CNB. In our recently reported series of 6,346 FNB,[12] only those 944 patients with suspicious (11%) or positive (4%) results were advised to have surgery. This 15% selection of patients is identical to that apparently found by Dr. Lo Gerfo in his CNB series.

Dr. Lo Gerfo mentions that the main risk of CNB is bleeding and describes 2 out of 3,000 patients who required operation within 24 hours of CNB. He believes that nonsurgeons "cannot handle the complications," and suggests that only physicians performing more than 50 CNB annually could ever be proficient. Could he outline what complications a nonsurgeon might expect if he or she was to "embrace this procedure" in his or her practice?

Response by Dr. Paul Lo Gerfo

This I believe shows what can be done by good people. Now the good people should try CNB and see how much better they can do.

Theoretically the problem is respiratory compromise, therefore the physician should have the facilities available to decompress the central compartment of the neck.

Discussion by Dr. James C. Sisson

Dr. Lo Gerfo makes a compelling case for coarse-needle biopsies in at least certain categories of patients with thyroid nodules. Two related questions arise. Does the injury inflicted on the thyroid and surrounding tissue (e.g., hematoma) by the coarse needle make subsequent thyroidectomy appreciably more difficult, especially if the excision of the lobe or entire thyroid is to be complete? Are there nodules, because of their clinical appearance (e.g., size), that do not require biopsy because they should be removed regardless of biopsy results? Is the surgical strategy affected by the preoperative biopsy results?

Response by Dr. Paul Lo Gerfo

Sometimes a large-needle biopsy will make the subsequent surgery more difficult, but most of the time it does not. It depends on the timing of operations.

There are nodules that are clinically malignant and require no biopsy.

Surgical strategy is based on the type of tumor, the age of the patient, anatomy of the parathyroids, whether there are nodules in the other lobe (at surgery), paraglandular node involvement, and whether the tumor is multicentric on cut section. The latter four of these are all intraoperative findings. I believe that for papillary carcinoma the needle biopsy diagnosis is very accurate, and if positive I do not need frozen section.

Discussion by Dr. Leonard Wartofsky

Of the small nodules that you have elected to excise as an office procedure, how many have been malignant? If there were any significant chance of malignancy, wouldn't such management not be highly recommended? And if there were little or no chance of malignancy, then why remove them?

What are the "other theoretical risks" that do exist but should not be seen by experienced physicians performing CNB?

Relative to CNB, it is significantly easier for most physicians to become skilled at FNB. Given the inherent difficulties in performing CNB, who should do them? Should one surgeon or thyroidologist develop this requisite skill? Would it then be preferable to concentrate performance of all CNB procedures in one medical center per large population area?

Is there any known or theoretical difference in the potential utility of CNB versus FNB for nodular thyroid glands in patients with a history of irradiation to the head and neck?

Does coarse-needle biopsy offer anything more than fine-needle biopsy in the patient with a history of radiation therapy to the upper body, assuming there is a nodule or a scan defect?

How often will your pathologist decline to make a diagnosis because of inadequate material?

Response by Dr. Paul Lo Gerfo

All patients are treated with a minimum of lobectomy and isthmectomy. Several patients with obvious papillary carcinoma done in the office had total thyroidectomy. Approximately 30% of patients are now undergoing thyroidectomy on an outpatient basis.

The theoretical risk is mainly respiratory obstruction secondary to bleeding.

FNB is easier and should be the main type of biopsy done. I was asked to write about CNB. I do many of them, and 50% are for patients for whom surgery had been recommended after FNB. I use FNBs all the time. They are great but not as accurate as CNB. CNB is never going to be popular and really should not become so—but it also should not be condemned. It has a role in patients with thyroid nodules just as FNB has a role.

I would not be able to biopsy a scan defect. There must be a palpable nodule. Otherwise, I think some of these irradiated people would be spared an operation if a coarse-needle biopsy were done.

About 1% or 2% of the time specimens are inadequate for diagnosis. In the beginning I had more inadequate specimens. But as I gained experience the number became less and less. You have to understand that I look at all of my biopsy specimens before I submit them to the pathologist, and also study the microscopic sections and discuss the findings with the pathologist.

Discussion by Dr. John T. Dunn

You suggest, from your experience at Columbia, that the FNB is no better than clinical methods for diagnosing thyroid cancer. This is contrary to the experience of many, and if your point is correct, no one should do needle aspirates. Since your statement has major implications for the general utility of FNB, it would be valuable for you to discuss the basis for your conclusions more thoroughly. The percent malignancy found at surgery may not be the best criterion for the value of FNB. Also, many factors could influence the decision for surgery other than the needle aspirate. In the "number recommended for surgery" this figure is also subject to bias, depending on the reassurance you convey to the patient about the value of the FNB versus the CNB.

Response by Dr. Paul Lo Gerfo

At our institution the incidence of cancer at surgery was 30% before FNB and 30% after FNB. This is *not* true of the other hospital we surveyed. Most major centers doing thyroid surgery have reported a cancer incidence of 30% to 36% without FNB. The incidence of cancer at surgery in most FNB series is less than 30%. I believe that a physician with a lot of experience will do well with or without FNB, but the problem of course is that there are few of these. Most endocrinologists see insufficient numbers of patients with thyroid disease to ever get really good at the evaluation, and FNB may help these people. Physician confidence in biopsy data may have an impact on surgical selection, but that would be true for all physicians, regardless of the biopsy method employed. In addition, approximately half of the biopsies I do are on people who have had an FNB and have been told they require surgery. I review these FNB slides and many are equivocal but some are inadequate. I still believe that the technique used is not the issue, but the issue is the physician using the technique.

Discussion by Dr. Mujtaba Husain

For the encapsulated microfollicular lesions we see at lobectomy, it is hard to tell whether there is invasion even though we take multiple sections. Therefore, it seems unlikely that there will be conclusive findings from the study of a 1-mm portion of the capsule on a large-needle biopsy.

I disagree with your statement that the larger tissue samples with coarse-needle biopsy will save many people from surgery. All you can establish from a cutting needle biopsy is the presence of a fragment of microfollicular tissue. Once this has been demonstrated, whether on a large or small tissue fragment, surgery is indicated and the amount of tissue doesn't make any difference.

I wonder if the multiple samples of fine-needle biopsy might not reduce sampling errors, compared to a single large-needle specimen.

Response by Dr. Paul Lo Gerfo

That is correct. I take them all out. I don't find frozen sections helpful in these tumors. If you have a lot of follicular lesions in the population of nodules with which you deal, since you will send most of them to surgery, and most of those lesions will be benign, that is a fundamental limitation to the prevalence of cancer you will have in your surgical specimens.

Some of the fine-needle specimens that show macro-microfollicular patterns will appear to be clear-cut benign hyperplastic nodules in large-needle specimens, and for them surgery may be avoided.

I usually take three large-needle samples from each nodule.

References

1. Ashcraft MW, Van Herle AJ: Management of thyroid nodules. Head Neck Surg 1981;2:216.
2. Lo Gerfo P, Starker P, Weber C, et al: Incidence of cancer in surgically treated nodules based on method of selection. Surgery 1985;98:1197.

3. Lo Gerfo P, Feind C, Weber C, Ting W: The incidence of carcinoma in encapsulated follicular thyroid lesions diagnosed by large needle biopsy. Surgery 1983;94:1008.
4. Hill L: Thyroid suppression. Arch Surg 1974;108:403.
5. Esselystyn DB Jr, Crile G Jr: Evaluation of various types of needle biopsies of the thyroid. World J Surg 1984;8:452.
6. Lo Gerfo P, Colacchio T, Weber C, Feind C: Comparison of fine needle and coarse needle biopsies in evaluating thyroid nodules. Surgery 1982;92:835.
7. Wang C, Vickery AL, Maloof F: Needle biopsy of the thyroid. Surg Gynecol Obstet 1976;143:365.
8. Hamburger JI, Husain M: Semiquantitative criteria for fine needle biopsy diagnosis: Reduced false-negative diagnoses. Diag Cytopathol 1988;4:14.
9. Gharib H, James EM, Charboneau JW, et al: Letter to the editor. N Engl J Med 1987;317:1664.
10. Morayati SJ, Freitas JE: The value of fine needle aspiration (FNA) in the management of thyroid nodules. J Nucl Med 1987;28:575.
11. Miller JM, Kini SR, Hamburger JI: In: Needle Biopsy of the Thyroid. New York: Praeger, 1983, pp 184–186.
12. Goellner JR, Gharib H, Grant CS, Johnson DA: Fine needle aspiration cytology of the thyroid, 1980 to 1986. Acta Cytol 1987;31:587.

10

Fine-Needle Biopsy: Extended Observations

JOEL I. HAMBURGER AND MUJTABA HUSAIN

The most efficient method for selection of thyroid nodules for excision or observation is needle biopsy, either fine-needle biopsy (FNB), which provides specimens for cytologic evaluation, or large-needle biopsy (LNB), which provides tissue segments or fragments for histologic evaluation. No other diagnostic method or combination of methods even approaches the sensitivity and specificity of needle biopsy. On the foregoing statement there is an overwhelming consensus. However, those may be the only points upon which there is general agreement. Controversies relative to the use of needle biopsy in the diagnosis of thyroid nodules may be outlined as follows:

1. Which procedure is preferable, FNB or LNB?
2. What is the best technique for FNB?
3. How can false negative FNB diagnoses be minimized?
4. Has needle biopsy eliminated the need for diagnostic ultrasound?
5. Should thyroid nodules be biopsied without prior thyroid imaging?
6. Is needle biopsy reliable in previously irradiated patients?
7. Should pregnant patients with thyroid nodules have needle biopsy?
8. How should patients with biopsy diagnoses of benign be managed?
9. Can biopsy data replace frozen-section data for surgical planning?

Which Procedure Is Preferable, FNB or LNB?

As shown by Lo Gerfo there is a place for LNB;[1] but it becomes a progressively smaller place as experience with FNB grows. Currently only about 5% of our patients have LNB procedures. The advantages of FNB are many and obvious. For the smaller nodules that constitute a majority of those detectable by physical examination (i.e., those smaller than 2 cm), only FNB is practical. These smaller nodules simply cannot accommodate the large needles used for LNB. FNB is clearly simpler and safer than LNB. The ease with which FNB can be performed permits more comprehensive sampling of larger nodules. Large cancers, especially follicular carcinomas, are often pleomorphic with some portions of the tumor composed of follicular elements that are virtually indistinguishable from benign follicular lesions. The major limitation of FNB is the inability to obtain adequate specimens from extensively degenerated nodules. For these tumors, a large-needle aspiration technique is often successful.

What Is the Best Technique for FNB?

Some say anesthesia is unnecessary because the anesthesia injection itself is as painful as the FNB.[2] How painful FNB is may depend upon whether one is performing the procedure or having it done. In our opinion at least six aspirates should be obtained from each nodule to assure adequate sampling. That much needling would try the forbearance of even the most compliant unanesthetized patient. Why not use local anesthesia? Some say the injected fluid obscures the margins of the nodule, making needle placement more difficult. This is nonsense. A few moments of gentle massage readily disperses the fluid restoring the anatomy.

Needles of various sizes from 27- to 18-gauge have been advocated for FNB. The finest needle practical is preferable because larger needles may induce excessive bleeding, diluting the specimens. Nevertheless, we do not like the 27-gauge needle because it is so flexible that it is hard to control during the insertion into the nodule. The 25-gauge 1½-inch needle is the best all-purpose needle for FNB. When we were learning FNB, we rather frequently used larger needles on very firm nodules from which we seemed unable to extract specimens with the 25-gauge needle. Subsequently we learned that a repetitive in-and-out movement of the needle through a distance of 1 to 2 mm within the nodule, as described by Lowhagen, is highly effective in producing the slight tissue disruption necessary to obtain FNB specimens.[3] If this maneuver fails, on rare occasions, it may be necessary to rotate the needle within the nodule through 360 degrees, permitting the sharp bevel of the needle tip to sever small tissue fragments for aspiration into the needle. When fluid first appears in the hub of the needle, suction is terminated and the needle is withdrawn.

It might seem unnecessary to comment upon where to insert the needle. To the novice it is obvious that the needle should be inserted into the most accessible point, the center of the nodule. For nodules of 1 cm or smaller, aiming for the center is the only practical approach. However, larger nodules often have undergone various degrees of degeneration. Degeneration is usually most extensive centrally. Aspiration of degenerated material will result in specimens too poor in follicular epithelium for reliable diagnosis. Proper technique is to insert the needles in a circumferential pattern at the periphery of the nodule where intact parenchyma is more likely to be found.

Some physicians use a syringe holder designed to permit one-handed performance of the biopsy, freeing the other hand to fix the nodule in position.[4] We found the apparatus clumsy to use, and it removed the biopsy hand some distance from the nodule with a loss of tactile sense that is important for FNB.

FNB specimens should be fixed immediately to prevent air drying, and stained with papanicolaou stain. This preserves the nuclear details upon which American cytopathologists place great reliance for cytologic diagnoses. European cytopathologists generally prefer air-dried Giemsa-stained preparations. They sacrifice nuclear detail for better cytoplasmic preservation. However, even Lowhagen has observed that for more difficult cases the papanicolaou preparations are superior.

How Can False Negative FNB Diagnoses Be Minimized?

False negative FNB diagnoses are reported in up to 10% of patients who have surgery after FNB diagnoses of benign.[5,6] Understandably this is a matter of concern to many physicians, some of whom remain reluctant to rely on FNB data for management decisions on patients with thyroid nodules. Some false-negative diagnoses simply reflect inadequate experience. Features of malignancy are just overlooked. Study of the illustrative and didactic material that is now available should curtail these elementary errors.[7-14]

Less obvious are the false-negative diagnostic errors made because diagnoses of benign have been forced upon too few cells. It should be obvious that more stringent criteria are needed to exclude diagnoses of cancer than to affirm such diagnoses. A diagnosis of cancer can sometimes be made on a single cluster of malignant cells. Malignant cells can only come from a malignant lesion. However, a similar or even larger number of benign cells may not be adequate to exclude cancer. After all, every thyroid gland has benign cells, regardless of whether a cancer is present, and those benign cells may be inadvertently aspirated while the malignant cells are missed. This may be the result of poor sampling, or because a pleomorphic tumor is sampled at the site of follicular structures that are indistinguishable from benign tissue. Hence adequate sampling is the prime prerequisite for the avoidance of false-negative FNB diagnoses. With this principle all would agree. On the matter of what constitutes adequate sampling there have been no specific guidelines, other than those from our practice.[14] Others have been content with one, two, or three aspirates.[4,15-17] We obtain at least six aspirates from each nodule, and insist that there must be at least six clusters of benign cells on each of at least two of the six smears prepared from those aspirates, and no malignant cells, before a diagnosis of benign is made. FNB specimens with lesser numbers of benign cells and, of course, acellular specimens are called inadequate to exclude malignancy.

Since 1985 we have employed the above criteria rigidly and 41 consecutive patients operated upon in spite of FNB diagnoses of benign had benign disease. An additional nine patients initially biopsied prior to 1985, and for whom rebiopsy later again revealed findings of benign disease, also had benign lesions at surgery. However, we have had nine recent patients with thyroid cancer (eight papillary, one follicular) who had lesser numbers of benign cells, and no malignant cells, on an initial FNB. Instead of diagnoses of benign, the specimens were called inadequate to exclude malignancy. Diagnoses of cancer were made on repeat FNB studies for four patients. Repeat FNBs were again inadequate for three others, and were not done on two patients.[18]

Using these strict criteria 21% of patients studied by FNB had unsatisfactory specimens; 15% having some benign cells, and 6% having acellular specimens.[18] This is a substantially larger portion of unsatisfactory FNBs than others report, although few centers provide data on this point, and none have offered definitions of what constitutes an unsatisfactory specimen. We attribute some of our failures to our willingness to try FNB on very small nodules (e.g., 0.5 cm and even smaller) if favorably placed for palpation. In addition we are receiving frequent referrals of patients for FNB after failure with FNB by others. In both of the foregoing situations a relatively high failure rate must be anticipated. But this is justified by the successes that are helpful in management decisions.

We are not concerned that FNB in our hands was able to provide usable data in only 79% of the patients. We prefer to recognize that FNB specimens are inadequate to exclude malignancy prospectively, rather than to learn about it months or years later when the diagnosis of cancer becomes clinically evident.

We inform all of our patients in advance that FNB sometimes fails to produce specimens adequate for the exclusion of cancer, even if the FNB is repeated. Should that be the case, we discuss with the patient the relative risks of observation and surgery, based on the available relevant clinical data, so the patient can participate in an informed decision on the subsequent management.

Has Needle Biopsy Eliminated the Need for Diagnostic Ultrasound?

We say yes. No ultrasound finding can either exclude or establish a diagnosis of cancer. Furthermore, no ultrasound finding eliminates the need for needling the lesion. If the lesion is cystic, needle aspiration may be diagnostic and therapeutic. Inspection of the fluid alone, should it be water clear, may make a diagnosis of a parathyroid cyst. Cytologic examination of the spun sediment from the cyst fluid infrequently provides evidence for malignancy.[19] About 50% of cystic lesions regress virtually completely in response to one or more aspirations. For solid lesions, FNB is employed to establish the probability of malignancy. Since regardless of the ultrasound findings needling will be done, and since needling will promptly differentiate cystic from solid nodules, as well as providing the additional anatomical information cited above, we agree with those who consider ultrasound an unnecessary and redundant expense.[20,21]

Should Thyroid Nodules Be Biopsied Without Prior Imaging?

In some centers enthusiasm for FNB is so great that it has become the initial and often the sole diagnostic procedure for thyroid nodules.[22,23] It has been claimed that this approach saves money by eliminating most thyroid images. In actual practice those savings may be illusory. Imaging before FNB would eliminate the need for FNB for the 5% to 10% of nodules that have increased function (e.g., autonomous function). Since the cost of FNB, including the pathologist's charge, is about twice that of thyroid imaging, unnecessary FNBs add disproportionately to costs. About 10% to 15% of the time, FNBs produce cellular specimens consistent with follicular lesions for which FNB findings cannot exclude malignancy. Some of these lesions will be autonomously functioning nodules.[24] Imaging will thus be needed if unnecessary surgery is to be avoided. FNB diagnoses of benign are obtained on about 70% of nodules. Most physicians advise thyroxine therapy and follow-up for most of these patients. FNB diagnoses of benign may be made on autonomously functioning nodules. Is it prudent to give thyroxine without imaging, and risk the induction of iatrogenic thyrox-toxicosis? Will patients who suffer that complication thank their physicians for trying to save money? We doubt it. To be sure, the thyroxine-treated patients can be watched for clinical and in vitro evidence of hyperthyroidism, and those who develop those findings can be imaged. Unfortunately for this strategy, patients treated with thyroxine frequently report symptoms compatible with hyperthyroidism, even when they are euthyroid. Also, elevated serum thyroxine values and subnormal supersensitive TSH assays are commonly encountered in seemingly euthyroid patients treated with thyroxine. The pressure to exclude underlying autonomous function with thyroid imaging will be hard to resist, and indeed it would be foolish to do so.

Finally, many physicians who obtain a FNB diagnosis of malignancy would want to have a preoperative image to assess the thyroid anatomy in advance of surgery, looking for other hypofunctional defects that may suggest multifocal disease. Preoperative images may be useful in some cases for comparison with postoperative images in the determination of whether radioactive iodine therapy might be indicated.

Thus, depending upon physicians' attitudes toward the above considerations, and the mix of nodule pathology encountered, the number of images ultimately avoided may be substantially less than some seem to suggest. It is even possible, when all is said and done, that FNB for all, followed by imaging when necessary, may be more costly than imaging for all, followed by needling when necessary.

Is Needle Biopsy Reliable in Previously Irradiated Patients?

It is well established that radiation therapy to the upper body, especially thymus, scalp, tonsils, and adenoids and for acne, is followed, usually 10 to 20 years later, by an increased probability of developing thyroid cancer. Concern for radiation-induced thyroid cancer has led some to take a very aggressive posture in favor of surgery for most, if not all, irradiated patients in whom thyroid nodules are detected, even if FNB diagnoses indicate benign disease.[6,25] However, it must be appreciated that exposure to radiation therapy increases the prevalence of both benign and malignant thyroid nodules. Indeed some believe that for any given thyroid nodule the chances of malignancy are not matcrially different whether there is a history of radiation therapy or not, and that irradiated patients can be selected for surgery or observation in the same way as nonirradiated patients.[26,27] This was one of the first issues we attempted to address when we started using needle biopsy for the diagnosis of thyroid nodules in 1976. Table 10.1 provides data on this point from two periods: first a period of somewhat less than 4 years between 1976 and 1980, and second a period of slightly more than 4 years between 1980 and 1984. In the earlier period we were influenced by histories of radiation therapy to advise surgery more aggressively. Thus, 39.8% of the irradiated patients were operated upon, compared to 25.2% of the nonirradiated patients. However, the percentage of patients studied by needle biopsy for whom malignancies were discovered at surgery was only slightly higher in the irradiated patients, at 10.8% compared to 9.4% for the nonirradiated. Impressed by the low yield of cancer when surgery was performed in spite of biopsy diagnoses of benign, in the second period we relied almost entirely upon biopsy findings to make recommendations for surgery or observation. Hence the proportion of patients operated upon was similar

TABLE 10.1. Prevalence of malignancy in nonirradiated and irradiated patients, 1976–1980 compared to 1980–1984.

	Nonirradiated		Irradiated	
	1976–1980	1980–1984	1976–1980	1980–1984
Number of nodules biopsied	1548	1654	176	181
Number of nodules excised (%)	390 (25.2)	438 (26.5)	70 (39.8)	44 (24.3)
Number of malignancies	146	213	19	20
% malignancy* in nodules biopsied	9.4	12.9	10.8	10.9
% malignancy in nodules biopsied combined series	11.2		10.9	

*Malignancy diagnosed on surgical specimens.

for the nonirradiated (26.5%) and the irradiated (24.3%). Again the percentage of patients studied by needle biopsy who had cancer discovered at surgery was similar for nonirradiated (12.9%) and irradiated (10.9%). The percentage of cancer in the two groups of irradiated patients was nearly identical (10.8% and 10.9%) even though the proportion of patients operated upon had been sharply reduced. Finally the percentages of malignancy found in surgical specimens for nonirradiated and irradiated patients, when the figures for the two periods were combined, were nearly identical, 11.2% (nonirradiated) and 10.9% (irradiated).

We believe that these data provide strong support for the concepts that the probability of malignancy for any given nodule is not materially different whether there is a history of radiation therapy or not, and that the irradiated patients can be evaluated and treated on the basis of the same criteria as the nonirradiated.[26,27]

Should Pregnant Patients with Thyroid Nodules Have Needle Biopsy?

We say yes. The procedure is exceedingly safe. If the patient has a benign nodule why should she have to worry about it for the balance of the pregnancy? On the other hand, if she has a nodule for which surgery is indicated, knowing about it promptly presents her with advantageous options. For example, she might wish to have the surgery in the second trimester, or even at the time of the delivery.[28] This would make it unnecessary for her to enter the hospital after the delivery (if the diagnosis were delayed until that time) and leave her newborn to the care of others. This is especially important for those who plan to breast-feed. Finally, a recent report suggests that nodules in pregnant women are more often malignant.[28]

How Should Patients with Biopsy Diagnoses of Benign Be Managed?

We emphasize to our patients that FNB diagnoses of benign are not guarantees, but only highly probable indications that the nodules biopsied are benign. We offer all such patients the option of a limited excision if the patient prefers the decisiveness of surgery and is willing to pay the price in terms of surgical risks and expense. Less than 10% of our patients select the surgical option. These patients are important to us since they serve as ongoing controls on which to establish the false-negative potential for FNB in our hands.

The majority prefer observation. We administer suppressive doses of levothyroxine unless there are contraindications, and we consider an age over 50 to 60 a relative contraindication unless there is associated hypothyroidism. In older people we are concerned that there may be non-suppressible function so that the addition of exogenous thyroid hormone may precipitate iatrogenic thyroxtoxicosis. Thyroxine treatment may be associated with substantial regression in up to 20% of nodules.[6] When that happens we are never certain that the regression was entirely due to suppression of thyrotropin, because needling of nodules alone may cause their regression. Nevertheless, for whatever reason, regression of nodules is reassuring evidence against malignancy.

Most nodules are unchanged after 1 or more years of treatment with thyroxine. Some consider this a reason to advise routine excision, regardless of the biopsy findings of benign.[6,25] We consider that too radical. Instead, we repeat the biopsy, and if the findings are again indicative of benign disease we are content to continue observation.[29]

What should be done when the nodule enlarges in spite of treatment with thyroxine? Does this mean cancer? In our experience, the most common cause is central hemorrhage. Other causes are inflammatory disease or unappreciated autonomous function. In total, those cases are much more likely than malignancy in a nodule for which a good-quality FNB has suggested benign disease. Imaging while thyroxine administration is continued (suppression imaging) will easily expose autonomous function. Needle biopsy will both diagnose and treat hemorrhage, and will identify inflammatory disease or cancer. Thus, there is no need to make a precipitate recommendation for surgery just because a nodule has enlarged during thyroxine therapy.

Can Needle Biopsy Data Replace Frozen-Section Data for Surgical Planning?

Once the decision to operate has been made, the surgeon must then decide how extensive an operation to perform. It is his or her hope that the initial operation will be the appropriate operation for the disease at hand.

For benign disease, and minimal thyroid cancer, lobectomy (usually including the isthmus) is generally considered adequate surgery. For larger and possibly more aggressive malignancies, a more extensive resection of the thyroid (e.g., total or near-total thyroidectomy) is preferred by most surgeons.

Traditionally, surgeons have placed reliance upon intraoperative frozen-section diagnoses for guidance in planning the extent of the operation. This strategy has well-recognized flaws. First, the patient cannot be told in advance how extensive an operation will be performed. This adds to the patient's anxiety and compounds the problem of assuring that informed consent is obtained for the operation. Perhaps more important is the realization that frozen-section diagnoses can be erroneous, commonly falsely negative and sometimes falsely positive, and probably most annoying is the frequent decision by the pathologist to defer the diagnosis until after study of the permanent section.[30,31] Consequently some patients are subjected to unnecessarily extensive operations, while many patients are faced with the prospect of a second operation after an inadequate primary operation.

The advent of needle biopsy suggested to us that it might be preferable to plan the extent of the surgical procedure on the basis of biopsy data. That approach would have several obvious advantages. Needle biopsy data are routinely obtained preoperatively in most clinics these days to select patients for surgery or observation. Needle biopsy data, with experience, are highly reliable. Basing surgical planning on needle biopsy data permits full preoperative disclosure of the surgical plan to the patient, and could save money by eliminating frozen-section procedures. To obtain these benefits, biopsy data must be at least as reliable, and preferably more reliable, than frozen-section data. This issue was studied in 359 patients operated upon after biopsy diagnoses.[30] Biopsy diagnoses fall naturally into three categories; cancer probability low, including diagnoses of benign nodular goiter, thyroiditis, follicular and Hurthle cell tumors without features suggesting malignancy; cancer probability high, including patients for whom biopsy diagnoses of definite malignancy or highly probable malignancy were made; and cancer probability intermediate, including patients with follicular (including Hurthle cell) lesions having biopsy features suggestive, but not diagnostic of malignancy. When needle biopsy suggested a low probability of malignancy (149 patients), a simple lobectomy would have been adequate surgery 90% of the time. For the remaining 10% the need for a second operation would have had to have been considered. Frozen-section diagnoses would have proved reliable for the same proportion of patients, 90%. However, frozen-section diagnoses were falsely negative in 6 of 109 cases, falsely positive in 4 of 8 cases, and deferred on 32 nodules, 5 of which were malignancies. Hence reliance of frozen section for surgical planning had the potential for both unnecessarily excessive surgery and inadequate surgery.

When biopsy diagnoses indicated a high probability of thyroid cancer (144 cases), malignancy was confirmed 135 (94%) times. Hence an appropriate cancer operation would have been correct with few exceptions. Frozen section was substantially less reliable. There were 18 (of 14) false-negative diagnoses, 2 (of 96) false-positive diagnoses, and diagnoses were deferred for 23 malignancies.

Neither needle biopsy nor frozen section was able to provide reliable guidance for suspicious follicular lesions. About 35% of them were cancers. For these patients, knowledge of the statistics should lead the surgeon to discuss the surgical options with the patient, explaining that a cancer operation might be unnecessarily excessive, but will avoid the necessity for a second procedure, while a limited operation will frequently be inadequate. Presented with these options, most of our patients prefer to have an initial cancer operation, rather than the less risky limited procedure. This preference is justified when the surgery is to be done by a qualified and experienced thyroid surgeon.

Summary

FNB is simpler and safer than LNB, is applicable to many more nodules than LNB, and with experience provides data of at least equivalent diagnostic utility. For best results with FNB at least six aspirates should be obtained from each nodule, circumferentially at the periphery of the nodule. Diagnoses of benign should only be made if at least two smears each contain at least six clusters of benign cells, and no malignant cells are found. Lesser numbers of benign cells may be insufficient to exclude malignancy. Although we do not use ultrasound, we prefer to image prior to FNB and do not biopsy autonomous nodules. Both previously irradiated and pregnant patients with nodules can benefit from FNB. After FNB diagnoses of benign, follow-up is necessary, and repeat FNB may be used for nodules that fail to respond to thyroxine. With experience FNB may provide more reliable guidance than frozen section for planning the extent of surgery.

Discussion by Dr. James C. Sisson

A word of caution should be expressed when considering a biopsy of a patient who may have hyperthyroidism. Hyperthyroid glands of Graves disease are very vascular, and the vascularity may surround an incidental and suspicious hypofunctioning nodule. A small hematoma in the thyroid gland is an inevitable consequence of multiple insertions of a needle, and the hematoma may be quite large in a vascular gland. Thyroid storm could be a consequence of the pressure from a large hematoma. The biopsy can await the euthyroid state.

Aspiration of cysts is often incomplete and even a small amount of residual fluid may obscure palpation of the solid part of the partially cystic nodule. A valuable role of ultrasound is to help determine the presence of a residual nodule after aspirating a cyst.

A biopsy can be performed during pregnancy, and thyroidectomy for a detected cancer can be performed during the pregnancy. But what is the hurry? I had one patient whose total thyroidectomy during the second trimester of pregnancy was performed by a highly skilled surgeon and still resulted in hypoparathyroidism that required considerable management to protect the fetus from hypo- and hypercalcemia.

Note should be made that thyroxine therapy for the benign hypofunctioning nodule has not been proven to be efficacious.[32]

Just as the results of aspiration biopsy may detect cancer missed on frozen sections, so also can frozen sections, especially those prepared and analyzed by experts, reveal cancer missed by aspiration biopsy. There is a place for both procedures when thyroid nodules are excised.

Response by Dr. Joel I. Hamburger

I agree with Dr. Sisson that biopsy of nodules in hyperthyroid patients can await restoration of euthyroidism.

I have not used ultrasound to determine the presence of a residual nodule (cystic, solid, or mixed) after aspiration of what appears to be predominantly a cystic nodule. If the aspiration fails to reduce the nodule to impalpability, I simply reaspirate. It is often possible to obtain cells from a persistent solid component of the nodule after the fluid has been withdrawn. If the nodule is impalpable, I don't think it is necessary to pursue the matter with ultrasound, and attempt, as some do, to introduce a needle into the nodule by ultrasound guidance. If any of the participants have done this, I would like to know how successful they have been.

Dealing with a possibly malignant nodule in a pregnant patient is certainly controversial. I have not had the experience of having to treat iatrogenic hypoparathyroidism in pregnancy, and I agree that I would rather not provide candidates for that unhappy event. However, knowing that there is a malignancy would permit the option of performing the surgery during the hospital stay in which the delivery is accomplished. This avoids the need for rehospitalization later and leaving the newborn in the care of others.

For many years physicians have treated thyroid nodules with thyroxine (LT_4), hoping to shrink them and thus reduce concern for malignancy. Although most nodules do not shrink on LT_4[6] for those that do the treatment is useful. Some consider lack of further growth also reassuring.[33] However, Gharib et al[32] say that LT_4 is no more effective than a placebo. In response to my query, they showed that in 6 months 5 of 25 nodules treated with a placebo shrank more than 50%, compared to 4 of 28 nodules treated with LT_4.[34] Lesser responses were observed in similar proportions of the two groups.[34] Interestingly, 8 nodules in the placebo group and 9 in the LT_4-treated group enlarged. The surprising placebo responses prompted me to compare size changes in all my LT_4-treated nodules with those in untreated nodules, all of which were followed for at least 1 year. Since only substantial shrinkage, easily detectable by palpation, is considered persuasive evidence against malignancy to most clinicians, ultrasound was not used. Of 100 consecutive LT_4-treated nodules, 19 shrank to impalpability, or by more than 75% leaving residuals of 0.5 cm or smaller; 9 shrank by 50%, with residual nodules of 3/4 cm or larger; and 72 exhibited no shrinkage to less than 50% shrinkage. No LT_4-treated nodule increased in size by more than 0.5 cm (the limits of accuracy by palpation).

Of 51 consecutive untreated nodules, 4 decreased in size to impalpability, 4 enlarged by 25% to 50%, and 43 exhibited negligible or no change in size.

Whether the longer duration of treatment, the larger number of nodules studied, or geography account for the differences in my patients from those of Gharib et al, perhaps more study of this issue is needed before the wisdom of those who preceded us is rejected.

The role of frozen section in patients who have had needle biopsy depends upon the reliability of FNB data and the skill with which frozen section is done, and this is something that each institution must determine for itself, as Dr. Sisson seems to suggest. Nevertheless, there are certain generalizations that may be made. FNB diagnoses of definite papillary carcinoma become almost 100% reliable with even modest experience. Intraoperative evaluation of those lesions by frozen section is clearly not cost effective. FNB diagnoses of suspicious follicular (including Hurthle cell) lesions are seldom decisive, and frozen-section diagnoses are similarly indecisive nearly always, with both false-negative and false-positive results possible. Until an institution shows that it can make reliable frozen-section diagnoses on follicular tumors (and I know of no such experience), the procedure may be best avoided. Perhaps the best use of frozen section is for the evaluation of a nodule discovered for the first time during the operation (especially when the lesion for which the surgery is performed had FNB findings of benign or only suspicious), or for the evaluation of enlarged lymph nodes that may harbor metastatic thyroid cancer.

Finally, frozen section may be useful to uncover false-negative FNB diagnoses when surgery is done in spite of FNB diagnoses of benign. The false-negative potential for FNB can easily be reduced to less than 5% if adequate criteria for adequacy of sampling are established. Therefore, the cost effectiveness of frozen section might be argued. Hence, although frozen section can be useful, its usefulness decreases as FNB expertise grows.

Discussion by Dr. Leonard Wartofsky

How similar were the nonirradiated patients to the irradiated patients with respect to other risk factors for thyroid cancer? That is, could the nonirradiated patients, by chance, have had a greater prevalence of other risk factors for malignancy?

I found your conclusion that previous radiation therapy may not be a significant risk factor for malignancy to be very interesting, and will look forward to the discussion of these data, particularly to comments by Dr. DeGroot.

Response by Dr. Joel I. Hamburger

This question was addressed in an earlier publication on this subject.[35] Our findings suggest that other clinical factors were more suspicious for cancer in the irradiated than the nonirradiated patients. For example, the female-to-male ratio for the irradiated patients was 1.6:1, whereas it was 4.4:1 for the nonirradiated. In nonirradiated patients, males were more likely to have cancer (43%) compared to females (36%). However, in the irradiated patients, the prevalence of cancer in males was only 22%, compared to 30% for females. Also, more irradiated than nonirradiated patients were younger than 40 years old. Hence, age and sex factors favored malignancy in the irradiated patients.

I did not say that previous radiation therapy is not a risk factor for malignancy. Surely it is. What I have said is that radiation increases the risk for both benign and malignant nodules, and that for any given nodule the chances of malignancy are not materially different whether there is a history of prior therapeutic radiation or not. In either case, the benign nodules outnumber the malignant ones by about 10 to 1. Since Dr. DeGroot has considerable experience on this subject, I will ask him to comment.

Discussion by Dr. Leslie J. DeGroot

I think the incidence of adenomas doubtless exceeds the incidence of carcinomas by more than 10-fold, although I am not sure it would be easy to prove that. If you will look at our 1983 paper,[36] in our group of patients referred only for a history of radiation therapy, the cancer incidence was 4.2% overall, while 53% of the patients undergoing operation, that is 13.7% of the total, had adenomas. However, usually tumors with a malignancy have it in the setting of innumerable small nodules in the thyroid. Thus, I think the actual frequency of benign lesions greatly exceeds that of malignant.

Discussion by Dr. John E. Freitas

Our technique for fine-needle aspiration of thyroid nodules differs somewhat from the technique reported. We use a 22-gauge 1½-inch needle attached to a 10-cc syringe containing 2 cc of air. Only three passes are made through the nodule with suction applied as the barrel of the syringe is rotated. The cellular material is expelled onto a glass slide without having to disconnect the needle from the syringe. Two slides are made from each pass for a total of six slides per nodule. The adequacy of each slide is assessed using the same criteria as discussed in the article. Our number of unsatisfactory specimens appears to be the same or less than that quoted using six passes with a 25-gauge needle. This technique has worked well for us over the past 10 years.

The importance of having a skilled, interested thyroid cytologist available to implement the fine-needle aspiration technique in the evaluation of thyroid nodules cannot be emphasized enough. There is a significant learning curve for thyroid cytology and this must be recognized by those who advocate needling all nodules without prior scanning. We have noted steady improvement in our specificity of fine-needle aspiration cytology over the past several years. The thyroidologist and thyroid cytologist need to maintain a close working relationship to achieve the best clinical results.

Response by Dr. Joel I. Hamburger

Before settling on the 25-gauge needle for fine-needle biopsy, we tried various needle sizes. We rejected the 22-gauge needle because we got unnecessarily excessive bleeding in many cases. This was without rotating the syringe barrel, which would produce more disruption and even greater bleeding. An additional objection that I would have to Dr. Freitas's technique of taking only three aspirates and making two slides from each aspirate is that three aspirates give one 50% of the sampling that is obtained with six aspirates. Depending upon the size of the nodule and the pleomorphism of the tissue, it is at least theoretically better to sample more sites. I would be interested to know if Dr. Freitas has done any systematic follow-up, particularly to assess the potential for false-negative diagnoses with his method.

Response by Dr. John E. Freitas

I am aware of only two false-negative fine-needle aspirations in our last 750 patients. However, since we are usually not providing long-term follow-up care for the majority of these referred patients, it is possible that there are other false-negative aspirations that we are not aware of despite yearly requests for follow-up to the referring physician.

Discussion by Dr. Paul Lo Gerfo

For the patients who get biopsies that are inadequate do you repeat the biopsy? What is the secondary failure rate?

Your numbers indicate that you send approximately 25% of your patients to surgery. This number is higher than most groups doing FNB (20%) and certainly higher than most of us doing coarse-needle biopsy (CNB) (15%)—why is this?

Your data show that 10% of all patients referred to you have cancer. Most old surgical series done at times when all nodules were excised indicate that cancer incidence in nodules is only 4%.[37] The number provided by Ascraft and Van Hurle[38] on 10,387 also shows an overall cancer incidence of 4.1%. In another 2,063 patients with CNB the incidence is 6.5%. The Framingham study shows an incidence of 3%. Why are your numbers so different? Are you calling microfollicular lesions cancer?

Response by Dr. Joel I. Hamburger

We repeat FNB studies when the initial FNB fails to produce adequate cells to exclude a diagnosis of cancer unless:

1. The patient refuses the repeat FNB.
2. The nodule shrinks substantially in a few months (with or without thyroxine therapy).

Table 10.2 gives our findings when an initial FNB produced some cells, but inadequate numbers to fulfill our criteria for reliable exclusion of malignancy, and when the initial FNB specimens contained no thyroid cells. In both cases repeat FNBs were sometimes effective both to demonstrate and to exclude malignancy.

When Dr. Lo Gerfo says we send approximately 25% of our patients to surgery, he is referring to obsolete data, not related to our current practice. In our experience between 1985 and 1987,[18] 158 (18%) of 888 nodules had surgery, and if we exclude 33 who had the operation in spite of benign FNB diagnoses, only 14% of patients had surgery advised by us because of FNB findings. Seventy (56%) of the 125 for whom we advised surgery had cancer.

In our most recent series (1985–1987) we identified cancer in 70 of 888 patients, a prevalence of 7.9%. This is less than we had in our earlier

TABLE 10.2. Repeat FNB diagnoses after initial diagnoses of inadequate or acellular.

		Repeat FNB diagnosis				
Initial FNB diagnosis	No. pts.	Acellular	Inadequate	Benign	Cellular ad.	Pap cancer
Inadequate	54	6	26	19	1	2
Acellular	27	11	6	7	1	2

patients, but more than the prevalences cited by Dr. Lo Gerfo from various sources. I think that the difference is accounted for by our restricting our analysis to solid, cold nodules. Other reports deal with patients who have cystic and functional nodules, and in many instances just diffuse goiters. We do not call microfollicular lesions cancers. The 70 cancers included 59 papillary, 5 follicular, 1 Hurthle cell, 2 metastatic cancers to the thyroid, 2 lymphomas, and 1 medullary carcinoma.

Probably a more important explanation for the higher prevalence of cancer in surgical specimens than the 3% to 6.5% cited by Dr. Lo Gerfo relates to the issue of percentage of what. The reports to which Dr. Lo Gerfo refers were dealing with the percentage of nodules that were malignant. Dr. Goellner found 5.8% malignancy in 6,300 nodules studied by FNB.[39] However, more than one FNB was performed on patients with more than a single nodule. My data, and that from most current reports, deal with the percentage of patients (not nodules) studied by FNB who had cancer.

Table 10.3 gives data from 10 centers published from 1980 and afterward. There were 4,818 patients studied, of whom 416 (9%) had malignant lesions. Included is the paper from Dr. Lo Gerfo's institution, the first author of which was Callachio, and their prevalence of cancer was 8%.

Discussion by Dr. Ian D. Hay

In contrast to Dr. Hamburger, we have at Mayo employed the Cameco syringe pistol and do not regularly give local anesthetic. Typically, we perform one to four aspirates and prepare one to 10 slides that are immediately fixed in 95% alcohol.

As recently reported by Goellner et al,[39] to be considered adequate for interpretation, a slide must contain five to six groups of well-preserved cells, with each group consisting of 10 or more cells. In a series of 6,346 specimens examined during 1980–86, 1,299 (20%) were found to be "nondiagnostic." Approximately one half of these unsatisfactory aspirates were due to cystic lesions, and on reaspiration approximately 60% of these cases had diagnostic cytologic results.

Although we do not routinely use sonography in the evaluation of nodular thyroid disease, we have found a role for high-resolution sonography in three circumstances. First, ultrasonographic-guided biopsy has been performed on nodules that had previously yielded nondiagnostic material by the traditional FNB technique. Second, we have used ultrasound to guide biopsy of impalpable neck masses in patients considered at high risk of locally recurrent thyroid cancer. Third, when preoperative imaging has been requested in thyroid cancer cases, we tend to use high-resolution sonography, rather than scintigraphy or CT scanning, to assess the extent of malignant disease and to document the presence of any associated suspicious cervical lymphadenopathy or parathyroid pathology.

TABLE 10.3. Cancer yield in thyroid nodules studied by FNB.

Author	Year	No. cases	No. excised (%)	No. malignant (%)*	% Malignancy in surgical specimens
Collacchio et al[40]	1980	300	80 (27)	23 (8)	29
Norton et al[41]	1982	102	21 (21)	10 (10)	48
Suen & Quenville[42]	1983	304	79 (26)	37 (12)	47
Hamaker et al[43]	1983	116	41 (35)	21 (18)	51
Gharib et al[17]	1984	1,970	364 (18)	158 (9)	43
Ramacciotti et al[6]	1984	221	126 (57)	25 (11)	20
Brauer & Silver[44]	1984	224	134 (60)	26 (12)	19
Boey et al[45]	1984	384	157 (41)	37 (10)	24
Silverman et al[46]	1986	309	60 (19)	8 (3)	13
Hamburger & Husain[18]	1987	888	159 (18)	71 (8)	45
		4,818		416 (9)	

*Percent of the total number evaluated for which malignancy was detected.

Like yourselves, we evaluate and treat the irradiated patient with a palpable nodule in the same manner as the nonirradiated. We have performed FNB on numerous pregnant women and, where a suspicious or positive result has been obtained, have permitted the option of second trimester surgery but have usually recommended holding off neck exploration until after delivery or later in the postpartum period. We now rarely administer T_4 suppressive treatment to patients with negative cytologic results, but prefer to reevaluate clinically and rebiopsy only if there has been significant interval change in the nodular lesion.

Response by Dr. Joel I. Hamburger

It is interesting to learn how different physicians using different techniques for FNB can still get good results. Your technique suggests that sometimes you make more than one smear from a single aspirate. This is only possible if the aspirate is more voluminous (dilute) than desirable. It might be useful to compare the cytologic content of such smears with that on smears of specimens that consist of a single drop of fluid. We seem to obtain more aspirates than anyone else. We do this because we want to be as sure as possible that we have enough cells, especially to exclude cancer.

Whether anesthesia is necessary depends upon two considerations. First is how many aspirates you take. Second is whether you are performing the biopsy or having it performed on you.

Criteria for adequacy of FNB samples depend upon whether one is attempting to confirm or exclude a diagnosis of cancer. A single cluster of cancer cells on one slide may be enough to suggest cancer, because only cancers have clearly malignant cells. To exclude cancer requires much more cytologic material, a lesson we learned by making false-negative diagnoses based on too few benign cells, or benign cells on only one aspirate. After all, every thyroid gland contains benign cells that can accidentally be sampled by FNB if the malignancy happens to be missed. Also some cancers (especially follicular) are highly pleomorphic so that cells from one area may be indistinguishable from benign cells, while cells from another area can readily be identified as suspicious for malignancy or frankly malignant. This is why we insist upon at lest six clusters of benign cells on at least two slides from separate aspirates before we make a diagnosis of benign. At the Washington, D.C. meeting of the American Thyroid Association (1987) we presented nine patients with thyroid cancer for whom FNB produced only benign cells, but lesser numbers than we considered adequate to exclude cancer. Because we appreciated that these specimens were inadequate to exclude cancer, we either repeated the FNBs or advised surgery. For two of these patients there were six clusters of benign-appearing cells on one slide. From this experience I would caution you that this is not enough material to exclude cancer.

In reviewing the article by Goellner et al,[39] to which you kindly referred me, I found that in 130 patients for whom surgery was performed in spite of FNB diagnoses of benign, 8 patients had cancer, 5 of which were papillary carcinomas. Dr. Goellner attributed the negative results in those papillary carcinomas to inadequate sampling. Since we have rigorously adhered to our criteria of at least six clusters of benign cells on at least two slides (not just one slide as Goellner et al accept), we have had no malignancies in about 60 patients (the latest tally) operated upon in spite of FNB diagnoses of benign. Hence, I think that Dr. Goellner's data offer support for our criteria for adequacy. I would be pleased if you or Dr. Goellner have any further comment on this point.

I read from time to time of the usefulness of ultrasonic guidance for FNB. I have not tried it. I would like to know how often a repeat FNB with ultrasound guidance provides adequate specimens when an initial FNB fails. Repeat FNBs without ultrasound were successful 32 of 81 times in our 1985–87 series. If repeat FNB with ultrasound guidance is better than that (and remember we require at least two slides, each having at least six clusters of cells), I would be convinced that it is worth trying, assuming that one has a reliable ultrasonographer. The other uses of ultrasound that you propose are interesting. It would be useful to have data on the results of such studies in a series of cases.

Whether one uses T_4 suppressive therapy for thyroid nodules depends upon one's experience. In the report by Gharib et al[34] from your institution, about 20% of nodules shrank substantially during thyroxine therapy. This is consistent with my experience and that of others. However, it was observed that an approximately equal proportion of nodules treated with a placebo also shrank substantially. This is not consistent with my experience, and I doubt it is consistent with yours. Do you really believe that 20% of all cold, solid nodules "treated" with a placebo will shrink by 50% or more? Since my untreated nodules do not shrink spontaneously nearly so often as my thyroxine-treated nodules, I use thyroxine.

Discussion By Dr. Ian D. Hay

Both Dr. Goellner and I are in agreement with you that "more stringent criteria are required to exclude diagnoses of cancer than to affirm such diagnoses." Currently each member of our thyroid "core group" will typically obtain two or three (and very rarely four) aspirates, in contrast to your obtaining "at least six." Apparently, you prepare only one smear from each aspirate, whereas often we will make more than one smear from a single aspirate. For adequate sampling, your cytopathologist insists on "at least six clusters of cells on each of at least two smears," whereas Dr. Goellner typically demands six clusters of well-preserved cells on one or more slides. Despite these differences, you have, since introducing your more strict criteria, obtained unsatisfactory specimens in only 21% of cases, while in our reported experience with 6,346 aspirates, we have a nondiagnostic rate that is only 1% different from yours. It is Dr. Goellner's opinion, which I share, that "more stringent criteria would perhaps decrease, but probably not totally eliminate the problem of false negatives!" It is his belief that we have a "very low false-negative rate" but admits in the light of your recent comments, that he will "probably read them (the smears) a little more strictly in the future!" I would agree with Dr. Goellner that the "nondiagnostic" and "suspicious for malignancy" cytologic results represent greater problems to us than the rare "false-negative" cases.

I do not accept that the paper by Gharib et al states that "20% of cold, solid nodules treated with a placebo will shrink by 50% or more." Of the 25 nodules "treated" by placebo, only 9 (36%) were purely solid. Fifty percent of the T_4-treated and 64% of the placebo-treated group had nodules with a cystic component. A reduction in size of $\geq$ 50% was obtained in only five of the placebo-treated and four of the T_4-treated patients. It is, I believe, noteworthy that six of the placebo-treated and four of the T_4-treated nodules were "predominantly cystic." I personally rarely attempt to suppress nodules with thyroxine, but I am certain that, like myself, Dr. Hamburger has observed cystic nodules shrinking spontaneously. Perhaps, in this series, some of the observed "substantial" shrinkage can be explained by "spontaneous" changes in the volume of fluid in the cystic component of the nodules in both groups.

Response by Dr. Joel I. Hamburger

Dr. Hay suggests that some of the observed "substantial" shrinkage can be explained by "spontaneous changes" in volume of fluid in the cystic component of the nodules in both groups. It seemed to me, and still does, that the possibility of "spontaneous" changes in the fluid volume of the cysts was less likely than changes induced by the needle punctures associated with FNB. I raised this issue[47] with Dr. Gharib, who stoutly denied that possibility, assuring me that the nodules were measured initially after FNB, and subsequently before a repeat FNB.[34] Although there could have been cystic areas that were not successfully punctured during FNB, or that reaccumulated fluid after the puncture and before the ultrasound measurement, and those cystic components could have "spontaneously" regressed, coincidentally in the same proportion of patients in the two groups, it seems to me that such an explanation might be difficult for most thyroidologists to accept. Gharib et al did not tell us if any of the nodules with cystic components were substantially smaller to palpation immediately after FNB (something that is commonly observed) and how many reenlarged because of fluid reaccumulation (i.e., repeat bleeding) soon after the FNB (something else that is commonly observed).

The potential implications of the large proportion of nodules with cystic components were not discussed by Gharib et al in their original report or in their response to my query. It might be reasonable to conclude that those authors did not think that the inclusion of those nodules made any difference, and it was merely by chance that similar proportions of the nodules in the placebo- and thyroxine-treated groups had cystic components of various volumes.

It seems clear that this is a fatal flaw in their study. The study should have been confined to solid nodules.

Worse yet, even sophisticated physicians who have read Gharib et al[32] may not appreciate this flaw, and conclude that: "thyroxine therapy for the benign hypofunctional nodule has not been proven to be efficacious" (see comment by Dr. Sisson, above).

Prompted by concern for the potential erroneous interpretation of the data of Gharib et al, I compared the responses to thyroxine for solid, cold nodules, with spontaneous changes in untreated nodules (see my response to Dr. Sisson's comment, above). Thyroxine-treated nodules were more than twice as likely to shrink substantially (i.e., by 75% to 100%) than untreated nodules. Hence I would like to know why Dr. Hay does not use thyroxine for at least a few years to try to obtain the benefits possible for this important subset of responders. Surely substantial regression, when it occurs, is of considerable reassurance to patient and physician.

Discussion by Dr. James C. Sisson

I enjoyed reading your new evidence in support of thyroxine treatment for thyroid nodules. On theoretical grounds, thyroxine therapy should cause regression of some nodules (i.e., the size and growth of some nodules are sustained, at least in part, by endogenous TSH). Therefore, if thyroxine is given judiciously, it should cause regression of some nodules and harm no patient. The question is then: how many nodules will respond?

Your data indicate that, during thyroxine therapy, about 19% shrank to 0.5 cm or less and another 9% shrank by over 50%, for a total of 28%. Without treatment, only 8% shrank appreciably. Given the numbers of patients involved, the results in the two groups appear to be significantly different. Probably a biostatistician could help you apply the best statistical analysis to your data. However, a major question must be answered before analysis: were the two groups of patients, treated and untreated, selected in the same manner? (Was there some special reason that certain patients went untreated so that a bias makes the two groups not comparable?) Although ideally a study of therapeutic efficacy should be double-blinded, one may still obtain valid data without the blinds if no bias can be shown in the compared groups.

Evaluating fewer patients than you, Gharib et al. demonstrated that about 20% of thyroxine-treated and a similar percent of placebo-treated patients had nodules that shrank more than 50%. As you pointed out, the numbers of patients were relatively small (25 and 28), and the treatment period was relatively brief. I have no experience in comparing treated and nontreated nodules. Nevertheless, I, too, do not believe that Gharib et al have published the definitive work on the treatment of thyroid nodules.

Response by Dr. Joel I. Hamburger

Patients were selected for treatment or nontreatment with thyroxine on the basis of age. Patients older than 50 were not treated; younger patients were. I doubt that age difference explains the difference in results. In particular, I find the shrinkage of 50% or more in 20% of the placebo-treated patients of Gharib et al to be unbelievable. However, I am willing to be persuaded if confirmatory reports surface from other centers. I will ask the participants for their comments on this point.

Response by Dr. John E. Freitas

I have not done a systematic analysis of the percentage of patients that demonstrate a 50% spontaneous regression in nodular size in 6 months of follow-up. Twenty percent appears to be excessive from my experience.

Discussion by Dr. John T. Dunn

Please give the data on which you base the choice of six separate aspirates for each nodule. Do you have comparisons with lesser numbers of aspirates (e.g., three, four, and five). Also do you have any comparative figures for using the 25- versus the 22-gauge needle? In the nine patients with inadequate fine-needle aspirations (FNAs) who had cancer, how inadequate were the aspirates, by your criteria?

I generally agree with your comments about ultrasound being of limited value. However, it is useful on occasion when there is a deep-seated nodule that is difficult to recognize by palpation, and the ultrasound can assure you that you are actually sampling the nodule in question.

Several comments on your discussion of imaging before FNA: 1) In addition to describing nodule function, imaging can also give an idea of the entire gland, particularly when deeply buried in a fat neck. A nonhomogenous uptake pattern increases the possibility of Hashimoto's or multinodular goiter; while these diagnoses, of course, do not exclude cancer, they do increase the statistical likelihood of a particular nodule being benign. 2) In our institution the cost of scanning is about equal to the total cost for FNA. 3) If 90% of nodules are cold, they will need needle aspiration anyway; why not do it first and avoid many unnecessary scans?

4) The concern about scanning before T_4 Rx to avoid hyperthyroidism seems unnecessarily complicated. The serum TSH can show fairly well whether the patient is hyperthyroid, or at least whether T_4 Rx is reasonable.

It is worth emphasizing a point that I know you endorse—that the cytologist's skill is the most critical factor in the successful use of the FNA. The cytologist must recognize the limitations of sampling and the impossibility of distinguishing a follicular adenoma from carcinoma particularly as FNA is used more widely and less critically.

In your discussion of suppressive doses of T_4, how do you define "suppressive"? I think it is extreme to demand that a patient with a FNA-benign nodule have the serum TSH suppressed, which for many (perhaps most?) people means making them hyperthyroid even though it is not clinically obvious.

Commentary by Dr. Joel I. Hamburger

With your permission, Dr. Dunn, I would like to expand your initial questions to include the following:

1. Why do we take a minimum of six aspirates from each nodule?
2. When do we take more than six aspirates?
3. Can one tell by gross inspection whether smears will contain thyroid cells?
4. When six aspirates produce at least two smears with at least six clusters each, how do we know that the aspirates containing those cells were not the first two, and the additional four aspirates were superfluous?
5. How often were aspirates after the fourth or sixth needed to fulfill our criteria for adequacy of sampling?

This issue of how many aspirates should be taken by FNB is one that has been ignored more often than not in reports on the subject. Most authors who trouble themselves to comment suggest one to three, or several, and none reported having taken more than four on a regular basis except myself and my former associates. Obviously the more aspirates one takes the more work is involved. Why do we do it? When we started our evaluation of FNB in 1976 we only took four aspirates. A dozen false-negative diagnoses, many of which were attributed to forcing diagnoses of benign on a few clusters of benign-appearing cells, led us to reevaluate our diagnostic criteria for adequacy of cellular material, particularly for the exclusion of malignancy. After 4 years' experience with some 2,000 FNBs we empirically established our current criteria: at least six clusters of benign cells on at least two slides, prepared from separate aspirates. It was a matter of simple logic to suggest that one should have benign cells on at least two aspirates from different parts of a nodule, to assure adequate sampling and to exclude the inadvertent sampling of normal tissue adjacent to the tumor. Since false-negative diagnoses had been made previously from three to four clusters of benign-appearing cells, six clusters seemed to offer an additional margin of safety.

Since 1985 we have been studying the efficacy of these criteria in the avoidance of false-negative diagnoses. To assess the false-negative rate it is necessary to operate on nodule patients in spite of FNB diagnoses of benign. This is something that we have been reluctant to advise; and it is advice that most of our patients would not be willing to accept. Nevertheless, a small proportion of our patients with FNB diagnoses of benign do prefer the diagnostic and therapeutic decisiveness of surgery, as well as the greater simplicity of subsequent follow-up. These patients, about 60 in number since 1985, provide us with the opportunity to evaluate the risk of false-negative diagnoses. None of these 60 patients had a malignant nodule. In contrast, the recent paper by Goellner et al[39] from the Mayo Clinic reported that 8 of 130 patients with FNB diagnoses of benign had cancers. They attributed the five failures in papillary carcinomas to inadequate sampling. They performed one to four aspirations and accepted a total of five to six clusters as adequate. In spite of their false-negative errors, they did not take the next step and suggest that one should insist upon two slides each with six clusters of benign cells before excluding cancer. Part of the reason there is so much resistance to performing six aspirations is a reluctance to use local anesthesia. Without it, most patients would object to more than a few needle punctures. The argument that the instillation of the anesthesia can obscure the nodule is clearly specious. A moment's massage readily disperses the fluid.

When I presented data at the Washington, D.C. Meeting of the American Thyroid Association (1987), showing how our proposed criteria permitted the avoidance of false negative diagnoses, I was questioned over and over about why I took so many aspirates. A reviewer of one of my recent papers said that a requirement for six aspirates was "excessive." He offered no data to support his opinion and chose to ignore mine to the contrary. I am expanding at some length on this issue of adequacy of FNB sampling because, in the final analysis, FNB will be judged by its potential for error. The purpose of FNB is to reduce the high rate of unnecessary surgery that occurs when the prescription for surgery is based on clinical criteria. Those with a vested interest in performing surgery are pleased to point to a false-negative rate of 5% to 10% for FNB as a justification for not using FNB or not relying on FNB data for management judgments. Those of us who have been impressed with the enormous cost-benefit implication of FNB should not allow sheer laziness to mitigate those potential benefits.

If it is granted that our criteria seem to have eliminated most of the potential for false-negative diagnoses, we must still address the subsidiary issue of how many aspirates are necessary to obtain at least two smears, each of which has at least six cell clusters. To answer this question, I asked Dr. Husain to report the number of cell clusters on each slide we prepared. Slides were numbered in the order in which the aspirates were obtained. The gross appearance of the smears was recorded, remarking on whether the aspirate seemed to consist of degeneration material (green, brown, or cheesy white), was thin, pale, and water rather than of a consistency similar to blood, and was voluminous, scanty, or bloody.

In 50 consecutive patients from whom specimens that fulfilled our criteria for adequacy of sampling, we found that the gross appearance of the specimen was unreliable for the prediction of cellular content. However, voluminous bloody or scanty aspirates seldom contained many thyroid cells. The first aspirate was about twice as likely to contain six clusters of cells as the fifth or sixth aspirates. The first two aspirates both contained six clusters 16 times. Three or four aspirates were needed 20 times; five or six aspirates were needed 10 times; and seven or eight aspirates were needed four times in order to obtain at least two slides each of which had six clusters of cells. Hence, about 75% of the time, four aspirates would prove adequate. Using six aspirates improved adequacy of sampling to more than 90%.

When do we take more than six aspirates? When an initial FNB with six aspirates produces inadequate specimens, on a second attempt we routinely take at least eight aspirates. We also take extra aspirates whenever the smears look grossly less promising than usual, especially when they are voluminous and bloody, or scanty. We do not like to discard these specimens because sometimes they contain useful material. We just do not count them.

The nine thyroid cancer patients who had inadequate FNB samples a total of 12 times (once six times or on three occasions twice) had zero to four cell clusters nine times, six clusters twice, and nine clusters once. The acellular specimens are easily recognized as inadequate. But those with some cells might not be reported by all cytologists as inadequate to exclude cancer.

We tested 27-, 25-, 22-gauge and in some cases larger needles before deciding that the 25-gauge 1½-inch needle is the ideal needle for FNB. The 27-gauge needle was too flexible for precise placement of the tip. The larger needles too often produced excessively bloody and voluminous specimens. Some use them because insertion of a 25-gauge needle followed by the application of negative pressure frequently fails to produce a specimen. However, using an in-and-out movement of the needle through 1 to 2 mm of nodular tissue will induce enough tissue disruption to secure fine specimens. Those who have used larger needles have found it necessary to filter voluminous specimens to eliminate the excess fluid that need not have been aspirated in the first place had they used a 5-gauge needle. For the same reason, others have tried to concentrate voluminous specimens by absorbing the excess fluid with filter paper. It is quicker and simpler just to take another sample, especially when the skin has been anesthetized.

Whether ultrasound is useful in the evaluation of thyroid nodules depends upon the use one makes of the procedure. Many authorities agree with Dr. Dunn that ultrasound guidance of the needle for FNB may be helpful for deep-seated nodules that are difficult to palpate. However, I have yet to see any data on a series of such cases to show how often that technique was effective. I would certainly allow that in centers with the best equipment and highly skilled physicians, that use of ultrasound may be considered. But ultrasound is used most often, by far, in the initial stages of the evaluation of thyroid nodules to determine whether these lesions are cystic, solid, or mixed cystic-solid. The tacit presumption is that cystic or predominantly cystic nodules can safely be considered benign. I submit that this notion is unsound, and furthermore that this use of ultrasound is not only a complete waste of money, but the data are unreliable when provided by ultrasonographers in private clinics and community hospitals where most patients are studied, at least in metropolitan Detroit.

Thyroid cancer occurs with about the same frequency in cystic nodules as solid nodules in my practice. Even the seemingly purely cystic lesion on occasion may have a focus of cancer involving the cyst wall. Furthermore, regardless of whether a nodule is cystic or solid, needle aspiration is indicated.

In the case of the cystic lesion, aspiration is done for therapeutic as well as diagnostic purposes. One to three aspirations will eliminate about 50% of cystic thyroid nodules. The character of the cyst fluid may also be diagnostic (e.g., watery clear fluid is virtually diagnostic of a parathyroid cyst); and just as I am writing this response, a diagnosis of papillary carcinoma was made by the detection of psammoma bodies and papillary fragments in fluid from a predominantly cystic thyroid nodule.

For solid nodules, FNB will be performed for diagnosis. Since needle aspiration will be performed in either case, and since needle aspiration will immediately show whether the nodule is cystic, solid, or mixed, ultrasound is a redundant waste of time and money that also serves to delay a decisive diagnosis.

Finally, impressed by frequent erroneous ultrasound diagnosis, I recently reviewed the records on 345 consecutive patients with cold thyroid nodules referred to me for evaluation in 1987. Ultrasound studies had been performed prior to referral on 76 (22%) of those patients. Since all were subjected to needle aspiration, the ultrasound studies were completely redundant. Not only were they redundant, but they failed to correctly determine the cystic, solid, or mixed nature of the nodules studied 44 times (58%), failing to locate the nodule altogether 20 times (even though the nodules were readily palpable), calling solid nodules cystic 15 times, cystic nodules solid one time, mislocating the nodule or locating only one of two nodules three times, and miscellaneous errors five times. Granted, this abysmally poor performance is not necessarily indicative of the best that ultrasound can do, but it is probably closely representative of what ultrasound actually does do in routine clinical practice.

I was the first physician in Michigan to publish experience with ultrasound, and was probably the first to abandon the procedure. I would encourage those primary care physicians who are detecting thyroid nodules and ordering ultrasound studies to do an immediate aspiration instead, using a 22-gauge 1½-inch needle and a 10- or 20-ml syringe (depending upon the size of the nodule). This is easy and safe, and I can assure them that they will do a better job of diagnosis than their ultrasonographer colleagues, with the added benefit that they will cure some of the cysts they encounter.

I have suggested imaging before thyroxine therapy not in order to detect hyperthyroidism. The supersensitive thyroid-stimulating hormone (STSH), with or without serum hormone assays, and the clinical evaluation should exclude that diagnosis. What I want to exclude is the 5% to 10% of nodules that are functioning autonomously, but without enough hormone secretion to produce hyperthyroidism. If these patients are not diagnosed before T_4 Rx, the T_4 will be additive to the autonomous nodular secretion and could induce iatrogenic thyrotoxicosis. To be sure, this may be recognized if the STSH value is undetectable on a follow-up visit; but then one could not differentiate between T_4 overdosage and the summation effect of exogenous hormone, plus endogenous autonomous hormone secretion. Inevitably imaging will be done on many of these patients or they will be mistreated. Incidentally, FNB including the pathology charges is about $200 for my patients. Imaging is only about half that expensive.

In treating nodule patients with T_4, I aim for low-normal STSH values. If the STSH value is fully suppressed, I reduce the T_4 dose. I agree with you completely on this point.

References

1. Lo Gerfo P, Starker P, Weber C, et al: Incidence of cancer in surgically treated thyroid nodules based on method of selection. Surgery 1985;98:1197.
2. Baskin HJ, Guarda LA: Influence of needle biopsy on management of thyroid nodules: Reasons to expand its use. South Med J 1987;80:702.
3. Lowhagen T, Willems J-S, Lundell C, et al: Aspiration biopsy cytology in diagnosis of thyroid cancer. World J Surg 1981;5:61.
4. Miller TR, Abele JS, Greenspan FS: Fine-needle aspiration biopsy in the management of thyroid nodules. West J Med 1981;134:198.
5. Harsoulis P, Leontsini M, Economou A, et al: Fine needle aspiration biopsy cytology in the diagnosis of thyroid cancer; Comparative study of 213 operated patients. Br J Surg 1986;73:461.
6. Ramacciotti CE, Pretorius HT, Chu EW, et al: Diagnostic accuracy and use of aspiration biopsy in the management of thyroid nodules. Arch Intern Med 1984;144:1169.
7. Hamburger JI, Miller JM, Kini SR: Clinical Pathological Evaluation of Thyroid Nodules Handbook and Atlas. Southfield, Private Publication, 1979.

8. Miller JM, Kini SR, Hamburger JI: Needle Biopsy of the Thyroid. New York: Praeger, 1983.
9. Kini SR, Miller JM, Hamburger JI, Smith-Purslow MJ: Cytopathology of follicular lesions of the thyroid gland. Diagn Cytopathol 1985;1:123.
10. Kini SR, Miller JM, Hamburger JI: Cytopathology of Hurthle cell lesions of the thyroid gland by fine needle aspiration. Acta Cytol 1981;25:647.
11. Kini SR, Miller JM, Hamburger JI: Problems in the cytologic diagnosis of the "cold" thyroid nodule in patients with lymphocytic thyroiditis. Acta Cytol 1981;25:506.
12. Kini SR, Miller JM, Hamburger JI, Smith MJ: Cytopathology of papillary carcinoma of the thyroid by fine needle aspiration. Acta Cytol 1980;24:511.
13. Kini SR, Miller JM, Hamburger JI, Smith MJ: Cytopathologic features of medullary carcinoma of the thyroid. Arch Pathol Lab Med 1984;108:156.
14. Hamburger JI, Hamburger SW: Fine needle biopsy of thyroid nodules: Avoiding the pitfalls. NY State J Med 1986;86:241.
15. Anderson JB, Webb AJ: Fine-needle aspiration biopsy and the diagnosis of thyroid cancer. Br J Surg 1987;74:292.
16. Schwartz AE, Nieburgs HE, Davies TF, et al: The place of fine needle biopsy in the diagnosis of nodules of the thyroid. Surg Gynecol Obstet 1982;155:54.
17. Gharib H, Goellner JR, Zinsmeister AR, et al: Fine-needle aspiration biopsy of the thyroid. Ann Intern Med 1984;101:25.
18. Hamburger JI, Husain M: Semiquantitative criteria for fine needle biopsy diagnosis reduced false negative errors. Diagn Cytopathol 1988;4:14.
19. Rosen IB, Provias JP, Walfish PG: Pathologic nature of cystic thyroid nodules selected for surgery by needle aspiration biopsy. Surgery 1986;100:606.
20. Goldfarb WB, Bigos TS, Eastman RC, et al: Needle biopsy in the assessment and management of hypofunctioning thyroid nodules. Am J Surg 1982;143:409.
21. Abdel-Nabi H, Falko JM, Olsen JO, Freimanis AK: Solitary cold thyroid nodule: Cost-ineffectiveness of ultrasonography. South Med J 1984;77:1146.
22. Reasner CA II, Isley WL: Thyroid scanning in patients with suspected thyroid diseases. Ann Intern Med 1985;102:209.
23. Solomon DH: Cost-effective analysis of the evaluation of thyroid nodule, pp 227–231. In: Van Herle AJ, moderator. The thyroid nodule. Ann Intern Med 1982;96:221.
24. Walfish PG, Strawbridge HTG, Rosen IB: Management implications from routine needle biopsy of hyperfunctioning thyroid nodules. Surgery 1985;98:1179.
25. Blum M: The diagnosis of the thyroid nodule using aspiration biopsy and cytology. Arch Intern Med 1984;144:1140.
26. Crile G, Esselstyn CB Jr, Hawk WA: Needle biopsy in the diagnosis of thyroid nodules appearing after radiation. N Engl J Med 1979;301:997.
27. Utiger RD: Is external irradiation a risk factor for thyroid disease and thyroid carcinoma? JAMA 1979;242:2702.
28. Rosen IB, Walfish PG: Pregnancy as a predisposing factor in thyroid neoplasia. Arch Surg 1986;121:1287.
29. Hamburger JI: Consistency of sequential needle biopsy findings for thyroid nodules. Arch Intern Med 1987;147:97.
30. Hamburger JI, Hamburger SW: Declining role of frozen section in surgical planning for thyroid nodules. Surgery 1985;98:307.

31. Keller MP, Crabbe MM, Norwood SH: Accuracy and significance of fine-needle aspiration and frozen section in determining the extent of thyroid resection. Surgery 1987;101:632.
32. Gharib H, James EM, Charboneau JW, et al: Suppressive therapy with levothyroxine for solitary thyroid nodules: A double-blind controlled clinical study. N Engl J Med 1987;317:70.
33. Watts NB: Letter to the editor. N Engl J Med 1987;317:1663.
34. Gharib H, James EM, Charboneau JW, et al: Letter to the editor. N Engl J Med 1987;317:1664.
35. Hamburger JI, Miller JM, Garcia M: Do all nodules appearing in patients subsequent to radiation therapy to the head and neck areas require excision? In: Hamburger JI, Miller JM, eds. Controversies in Clinical Thyroidology. New York: Springer-Verlag, 1981, pp 237–266.
36. DeGroot LJ, Reilly M, Pinnameneni K, Refetoff S: Retrospective study of radiation-induced thyroid disease. Am J Med 1983;74:852.
37. Hill LD, Beebe HG: Thyroid suppression. Arch Surg 1974;108:403.
38. Ascraft MW, Van Herle AJ: Management of thyroid nodules. Head and Neck Surg 1981;3:297.
39. Goellner JR, Gharib H, Grant CS, Johnson DA: Fine needle aspiration cytology of the thyroid, 1980 to 1986. Acta Cytol 1987;31:587.
40. Colacchio TA, Lo Gerfo P, Feind CR: Fine needle cytologic diagnosis of thyroid nodules. Review and report of 300 cases. Am J Surg 1980;140:568.
41. Norton LW, Wangensteen SL, Davis JR, et al: Utility of thyroid aspiration biopsy. Surgery 1982;92:700.
42. Suen KC, Quenville NF: Fine needle aspiration biopsy of the thyroid gland: A study of 304 cases. J Clin Pathol 1983;36:1036.
43. Hamaker RC, Singer MI, DeRossi RV, Shockley WW: Role of needle biopsy in thyroid nodules. Arch Otolaryngol 1983;109:225.
44. Brauer RJ, Silver CE: Needle aspiration biopsy of thyroid nodules. Laryngoscope 1984;94:438.
45. Boey J, Hsu C, Collins RJ, Wong J: A prospective controlled study of fine-needle aspiration and tru-cut biopsy of dominant thyroid nodules. World J Surg 1984;8:458.
46. Silverman JF, Lee West R, Larkin EW, et al: The role of fine-needle aspiration biopsy in the rapid diagnosis and management of thyroid neoplasm. Cancer 1986;57:1164.
47. Hamburger JI: Letter to the editor. N Engl J Med 1987;26:1164.

Section Summary: Needle Biopsy Diagnosis of Thyroid Nodules

JOEL I. HAMBURGER

Why Do I Prefer Fine-Needle Biopsy (FNB) as the Principal Biopsy Procedure?

It is simpler and safer than coarse-needle biopsy (CNB). It is easier to obtain multiple samples with fine-needle biopsy (FNB) and this may be important for larger pleomorphic follicular tumors. With either technique, it is possible to improve the selection process so that the prevalence of cancer in excised nodules exceeds 50%, rather than the 20% to 30% prevalence that was the best that could be achieved prior to needle biopsy. As lo Gerfo said: "The technique used is not the issue as much as the physician using the technique." Since it is possible to do as well with FNB as with CNB, it would seem preferable to put effort in improving FNB technique, rather than trying to cover FNB failures by learning to use CNB. Perhaps the single biggest inadequacy in the use of FNB is the refusal to perform enough aspirations to assure adequate sampling. When too few aspirates are taken, the cytopathologist will have fewer cells to study and may yield to the temptation to force a diagnosis upon inadequate specimens. In particular, the problem is a false-negative diagnosis made on the basis of a few clumps of benign-appearing cells on a single slide. These errors were committed by my first cytopathologist before we knew better, and by the Mayo Clinic cytopathologists, and no doubt by many others. False-negative errors can be minimized by the simple expedient of taking more samples, at least six aspirates from each nodule, and insisting upon multiple clumps of benign cells on at least two of the slides before excluding malignancy.

Does a History of Radiation Therapy Invalidate the Use of Needle Biopsy?

There is lingering concern that prior radiation therapy to the upper body so predisposes to the development of thyroid cancer that the appearance of a nodule per se might be an indication for surgery. In my practice, about 20% of irradiated patients had thyroid abnormalities detectable by palpation, but only 2% had thyroid cancer. The data I have presented in this volume serve to support those who maintain that for any given thyroid nodule the chances of cancer are no different whether or not there is a history of radiation therapy. Hence there is no reason not to rely upon FNB in the decision of whether to operate or observe.

Should Ultrasound Be Abandoned in the Evaluation of Thyroid Nodules?

The use of ultrasound to assist in the selection of nodules for surgery or observation is based upon the faulty premise that cystic nodules are nearly always benign. Cystic cancers and cysts containing cancer are rather common. No ultrasound finding is decisive in the diagnosis of thyroid cancer, and no ultrasound finding precludes the necessity for needle biopsy. The initial needle aspiration immediately determines whether a nodule is cystic, partly cystic, or solid. If cystic, the aspirated fluid can be examined for malignant cells or psammoma bodies, or for parathormone if the fluid is the crystal clear to opalescent fluid suggestive of a parathyroid cyst. Aspiration may eliminate cystic lesions about half the time. For solid or partially solid lesions, needle biopsy will be performed. Ultrasound is only an expensive redundant procedure that delays the more decisive needle procedure.

The use of ultrasound to direct needle placement during FNB on small nodules or small solid portions of predominantly cystic lesions, although frequently advocated, has yet to be subjected to scientific evaluation. It may be appropriate in a research setting, but probably has at best a minor role to play in clinical practice. I say this because if the nodule is palpable, ultrasound guidance is unnecessary. If the nodule is impalpable (e.g., there is no palpable remnant after aspiration of a predominantly cystic lesion), I would be inclined to observe the patient. The cystic lesions with small malignant foci that I have detected have made their presence known by repeated refilling of the cyst after aspirations. I believe one can safely defer the detection of other impalpable cancers until they grow to a size at which they become palpable.

Should Benign Nodules Be Treated with Thyroxine?

Perhaps the most controversial issue that surfaced during the discussions of papers on needle biopsy had to do not with needle biopsy, but rather with the follow-up of nodule patients after a needle biopsy diagnosis of benign. Should they be treated with thyroid hormone or not? This controversy was brought into sharp focus by Gharib's paper from the Mayo Clinic that seemed to suggest that patients with thyroid nodules treated with thyroid hormone achieved a reduction in nodule size no more often than those treated with a placebo (i.e., untreated). To those of us who see large numbers of patients with thyroid nodules, it was not news that most of the nodules in patients treated with thyroid hormone did not shrink substantially. We did not need serial ultrasound studies to prove this. Indeed, since most of the nodules we treated with thyroxine were cold on thyroid imaging, one might infer that they were not responsive to thyroid-stimulating hormone (TSH), and thus should not respond to thyroxine because thyroxine could work only by suppressing TSH. Is that conclusion justified? Not necessarily. In the first place, imaging does not give decisive information on the absolute function of a nodule, only information as to the relative function of the nodule compared to the function of extranodular thyroid tissue. Low-level functional capacity of a nodule may be obscured by the much better function of the extranodular tissue. This, of course, is why it is helpful to excise all normal tissue before imaging will show that residual thyroid cancer can concentrate radioactive iodine. Before surgery, the cancer characteristically appears unable to take up the tracer, presenting as a cold nodule. However, responsiveness to TSH is often seen in terms of tracer uptake after thyroidectomy. It is not too difficult, therefore, to infer that benign nodules composed of thyroid parenchymal tissue may be responsive to TSH suppression by thyroxine.

Also, TSH has the capacity to stimulate both thyroid function and also the growth of thyroid tissue. Tissue that is less responsive to the function-stimulating properties of TSH may still be responsive to the growth-stimulating effect, an effect that may be suppressed by thyroxine.

Theoretical considerations aside, it would seem imprudent to discard the experience of so many physicians over so many years that indicated that some nodules do shrink substantially and even become impalpable during thyroxine therapy, simply because of a single study from the Mayo Clinic on a small number of patients studied for a short period of time, especially when that study seems to have been subject to major criticisms, as already discussed.

I continue to use thyroxine for nodule patients younger than about 50 years of age because those fortunate enough to respond to the treatment are less concerned for possible cancer. The treatment may still prevent further growth even if the nodule does not shrink. On the contrary, growth of the nodule during thyroxine therapy suggests the need for further study to exclude cancer. A final benefit from thyroxine therapy is that it serves to remind the patient that there is a continuing problem that needs periodic evaluation. I have seldom advised thyroxine to patients older than 50 (unless they were hypothyroid) because the coincidental presence of autonomously functioning tissue seems more common in older patients. I do not wish to induce iatrogenic thyrotoxicosis on the basis of the additive impact of exogenous thyroxine to the nonsuppressible hormone output of any autonomous tissue.

Index

MIX
Papier aus verantwortungsvollen Quellen
Paper from responsible sources
FSC® C105338

If you have any concerns about our products,
you can contact us on
ProductSafety@springernature.com

In case Publisher is established outside the EU,
the EU authorized representative is:
Springer Nature Customer Service Center GmbH
Europaplatz 3, 69115 Heidelberg, Germany

Printed by Libri Plureos GmbH
in Hamburg, Germany